Advances in 21st Century Human Settlements

Indexed by SCOPUS

This Series focuses on the entire spectrum of human settlements — from rural to urban, in different regions of the world, with questions such as: What factors cause and guide the process of change in human settlements from rural to urban in character, from hamlets and villages to towns, cities and megacities? Is this process different across time and space, how and why? Is there a future for rural life? Is it possible or not to have industrial development in rural settlements, and how? Why does 'urban shrinkage' occur? Are the rural areas urbanizing or is that urban areas are undergoing 'ruralisation' (in form of underserviced slums)? What are the challenges faced by 'mega urban regions', and how they can be/are being addressed? What drives economic dynamism in human settlements? Is the urban-based economic growth paradigm the only answer to the quest for sustainable development, or is there an urgent need to balance between economic growth on one hand and ecosystem restoration and conservation on the other — for the future sustainability of human habitats? How and what new technology is helping to achieve sustainable development in human settlements? What sort of changes in the current planning, management and governance of human settlements are needed to face the changing environment including the climate and increasing disaster risks? What is the uniqueness of the new 'socio-cultural spaces' that emerge in human settlements, and how they change over time? As rural settlements become urban, are the new 'urban spaces' resulting in the loss of rural life and 'socio-cultural spaces'? What is leading the preservation of rural 'socio-cultural spaces' within the urbanizing world, and how? What is the emerging nature of the rural-urban interface, and what factors influence it? What are the emerging perspectives that help understand the human-environment-culture complex through the study of human settlements and the related ecosystems, and how do they transform our understanding of cultural landscapes and 'waterscapes' in the 21st Century? What else is and/or likely to be new vis-à-vis human settlements — now and in the future? The Series, therefore, welcomes contributions with fresh cognitive perspectives to understand the new and emerging realities of the 21st Century human settlements. Such perspectives will include a multidisciplinary analysis, constituting of the demographic, spatio-economic, environmental, technological, and planning, management and governance lenses.

If you are interested in submitting a proposal for this series, please contact the Series Editor, or the Publishing Editor:

Bharat Dahiya (bharatdahiya@gmail.com) or
Loyola D'Silva (loyola.dsilva@springer.com)

Mahesti Okitasari · Paul Hudson ·
Masachika Suzuki
Editors

Equitable and Sustainable Cities

Economics and Disruptive Technologies

Springer

Editors
Mahesti Okitasari
United Nations University Institute for the Advanced Study of Sustainability
Tokyo, Japan

Paul Hudson
Department of Environment and Geography
University of York
York, UK

Masachika Suzuki
Graduate School of Global Environmental Studies
Sophia University
Tokyo, Japan

ISSN 2198-2546 ISSN 2198-2554 (electronic)
Advances in 21st Century Human Settlements
ISBN 978-981-92-2792-1 ISBN 978-981-92-2793-8 (eBook)
https://doi.org/10.1007/978-981-92-2793-8

This work was supported by No Funder.

This Springer imprint is published by the registered company Springer Nature Singapore Pte Ltd.
The registered company address is: 152 Beach Road, #21-01/04 Gateway East, Singapore 189721, Singapore

This book is dedicated to everyone working hard to build equitable and sustainable cities for our children.

Series Editor's Foreword

It is with great pleasure that I introduce this timely anthology, *Equitable and Sustainable Cities: Economics and Disruptive Technologies*, co-edited by Mahesti Okitasari, Paul Hudson, and Masachika Suzuki. This book emerges as our world grapples with unprecedented and incessant urban growth and the escalating impacts of climate change, placing immense pressure on cities to evolve sustainably and equitably. The urgency is underscored by the United Nations' Sustainable Development Goals (SDGs)—particularly SDG 11 (United Nations 2025), which positions cities and human settlements at the core of global policy goals, recognising their disproportionate environmental, social, and economic impacts in the 21st century.

This book is particularly significant because it directly addresses the multifaceted challenges of rapid urbanisation, including exacerbated intra-urban inequalities, urban poverty, and increased vulnerability to natural disasters (Dahiya et al. 2023; Dahiya and Das 2020; Das and Dahiya 2020; ESCAP and UN-Habitat 2023; Haase et al. 2018; UN-Habitat 2018, 2020, 2022, 2024; United Nations Department of Economic and Social Affairs 2025). It fills a critical research gap in current literature that is related to the critical and practical examination of whether urban innovation-centred financing and technologies achieve the sustainability goals they have set, including economic growth, social inclusion, and environmental stewardship. Scholars have sought to identify key novel finance (Zhang et al. 2024) and disruptive technologies (Radu 2020) pivotal for sustainable cities, yet that work leaves the links between these innovations and the distribution of economic and social outcomes over time and space largely unexplored (see Kemeny et al. 2025), especially with a specific focus on cities and regions. This book fosters such interaction by examining the relationship between innovations and inequality. As such, it responds to the need for a greater focus on social justice in the urban innovation scholarship in order to integrate a broader conception of equity and link innovation with better distributional outcomes (Lee 2025). The book also offers a comprehensive exploration of innovative and transformative approaches to city planning, governance, and urban resource management. The team of co-editors does so by bringing together diverse perspectives across a range of academic disciplines, allowing the volume to contribute to ongoing debates about how societies can foster urban environments that are not only resilient but also socially just.

The overarching purpose of this anthology is to explore multiple pathways towards achieving equitable and sustainable cities by examining how novel technologies, social innovations, and institutional frameworks can drive meaningful change. The overall context of the book is set out in the Chapter "An Introduction to Equitable and Sustainable Cities—Economics and Disruptive Technologies" authored by the co-editors. The core themes woven throughout the various chapters provide insightful analyses, highlight practical solutions, and present a critical perspective on next steps.

- *Innovative Urban Solutions.* This theme is brilliantly exemplified by Chapter "The Role of Patios in Creating Equitable and Sustainable Cities" by Delgado-Serrano, Redaelli, and Piedrahita. The chapter delves into retro-innovations like Cordoba's patios for passive cooling and social well-being, demonstrating how traditional wisdom can serve as a disruptive technology in fighting climate change and boosting urban equality.
- *Addressing Socio-Economic Disparities.* The book explores the intricate issue of energy poverty in developed nations, such as Japan, in Chapter, "Addressing Energy Poverty in the Developed Countries: A Case Study from Japan" by Zhu, Ramadhani, and Suzuki, highlighting the link between energy affordability and socio-economic factors. Furthermore, Chapter "How Innovative Financial Technologies Can Foster Equity and Sustainability in Cities and Regions" by Adi and Aljunaidy, examines how financial technologies (Fintech) can extend financial services fostering equity and bridging financial gaps within cities and regions if correctly managed.
- *Community-Driven Resilience.* The volume demonstrates the power of community-led initiatives, with Chapter "Housing Cooperative: A Force to Be Harnessed for Climate Change Mitigation and Adaptation" by Akanji and Hudson, highlighting housing cooperatives in Nigeria in promoting sustainable building practices, energy efficiency, and collective climate resilience. This theme is further explored in Chapter "NIMBY to YIMBY: Towards a Better Neighbourhood" by Li, Ren, and Chiwosi, which presents a compelling case study illustrating the transition from "Not In My Backyard" (NIMBY) resistance to "Yes In My Backyard" (YIMBY) acceptance in municipal solid waste management, fostering equitable urban sustainability through aligned policy, technology, and community agency.
- *Sustainable Finance and Energy Transitions.* This vital area is addressed in Chapter "Green and Sustainable Municipal Bonds: Accelerating Infrastructure Investment for Equitable and Sustainable Cities" by Liu, Okitasari, and Lau,

which investigates the transformative potential of green and sustainable municipal bonds in accelerating infrastructure investment for equitable and sustainable urban development. Additionally, Chapter "Current Status and Barriers for the Diffusion of Distributed Rooftop Photovoltaics in Southeast Asia: Lessons Learned from Germany, Rwanda, and the United States" by Zhu and Suzuki, analyses the diffusion of distributed rooftop photovoltaics in Southeast Asia, emphasising their role in poverty alleviation and energy security, alongside policy recommendations. Further enriching this theme, Chapter "Investing in Co-Benefits: A Financial-Economic and Political Explanation" by Van der Heijden, Rambharos, and Dorst, explores novel financial strategies and governance mechanisms for financing integrated spatial development crucial for sustainable cities, highlighting the complex interplay between financial arguments and political decision-making.

- *Integrating Diverse Knowledge Systems.* A unique contribution lies in the exploration of harmonising Indigenous and Local Knowledge (ILK) with modern science and technologies for Climate Smart Agriculture in Chapter "Harmonising Indigenous and Local Knowledge in Climate Smart Agriculture: Practical Insights from the Satoyama Initiative" by Nishi, aiming for a "triple-win" of agricultural productivity, climate resilience, and mitigation.
- *Enhancing Resilience through Technology.* Finally, the volume examines the pivotal role of insurance and Insurtech (technological innovations in the insurance sector) in strengthening natural hazard and climate resilience through improved risk assessment and customer engagement in Chapter "The Role of Insurance and Insurtech in Enhancing Natural Hazard Resilience" by Hudson.

Through these diverse thematic issues and case studies, this book offers valuable insights for policymakers, urban planners, financial institutions, researchers, and communities striving to build a more sustainable and equitable urban future going forward. The chapters presented in the edited volume have illuminated how innovative tools like Fintech and financing instruments and technological innovations can address complex urban challenges while ensuring climate action, human development, and social justice are in place. The authors have also highlighted the persistent socio-economic disparities taking place in the implementation of novel financing and disruptive technologies, signalling the need for further work to better understand solutions and enabling environment to mitigate these challenges.

Mahesti Okitasari, Paul Hudson, and Masachika Suzuki have done excellent and commendable work in conceptualising this anthology and bringing together this rich collection of well-researched papers. This co-edited volume stands as a testament to the power of interdisciplinary collaboration on a topical theme that weaves together economic analyses, technological innovations, and societal considerations in the rapidly urbanising world. A lot more needs to be done in this important area of interdisciplinary research for the benefit of humans and all sentient beings for whom Planet Earth is the home.

Bharat Dahiya
Director, Research Center for
Sustainable Development
and Innovation
School of Global Studies
Thammasat University
Pathum Thani, Thailand

References

Dahiya, B., & Das, A. (2020). New Urban Agenda in Asia-Pacific: Governance for Sustainable and Inclusive Cities (pp. 3–36). https://doi.org/10.1007/978-981-13-6709-0_1

Dahiya, B., De Pascale, F., De Pietro, O., Farabollini, P., Lugeri, F. R., & Mercatanti, L. (2023). Disaster Resilience and Human Settlements in the Anthropocene (pp. 3–18). https://doi.org/10.1007/978-981-99-2248-2_1

Das, A., & Dahiya, B. (2020). Towards Inclusive Urban Governance and Planning: Emerging Trends and Future Trajectories (pp. 353–384). https://doi.org/10.1007/978-981-13-6709-0_13

ESCAP, & UN-Habitat. (2023). Crisis Resilient Urban Futures: The Future of Asian and Pacific Cities 2023

Haase, D., Güneralp, B., Dahiya, B., Bai, X., & Elmqvist, T. (2018). Global Urbanization. In Urban Planet (pp. 19–44). Cambridge University Press. https://doi.org/10.1017/9781316647554.003

Kemeny, T., Petralia, S., & Storper, M. (2025). Disruptive innovation and spatial inequality. Regional Studies, 59(1). https://doi.org/10.1080/00343404.2022.2076824

Lee, N. (2025). Inclusive innovation in cities: from buzzword to policy. Regional Studies, 59(1). https://doi.org/10.1080/00343404.2023.2168637

Radu, L.-D. (2020). Disruptive Technologies in Smart Cities: A Survey on Current Trends and Challenges. Smart Cities, 3(3), 1022–1038. https://doi.org/10.3390/smartcities3030051

UN-Habitat. (2018). The State of African Cities 2018: The geography of African investment. Nairobi

UN-Habitat. (2020). World Cities Report 2022: The Value of Sustainable Urbanization. Nairobi

UN-Habitat. (2022). World Cities Report 2022: Envisaging the Future of Cities. Nairobi

UN-Habitat. (2024). World Cities Report 2024: Cities and Climate Action. Nairobi

United Nations. (2025). Sustainable development goal 11: make cities and human settlements inclusive, safe, resilient and sustainable

United Nations Department of Economic and Social Affairs. (2025). The Sustainable Development Goals Report 2025. New York

Zhang, W., Wang, J., & Jin, H. (2024). Digital finance, innovation transformation, and resilient city growth. Scientific Reports, 14(1), 7056. https://doi.org/10.1038/s41598-024-56998-z

Acknowledgements

We are deeply grateful to the many individuals whose contributions have been essential in bringing this edited book, *Equitable and Sustainable Cities: Economics and Disruptive Technologies*, to fruition.

First and foremost, the Editors would like to thank the contributing authors of the various chapters included in this volume: Mohamad Nadim Adi, Ajibola Anthony Akanji, Mais M. Aljunaidy, Blessings Chiwosi, Catalina Cruz Piedrahita, Maria Mar Delgado-Serrano, Hade Dorst, Eric Lau, Li Li, Wenjie Liu, Maiko Nishi, Deanty Mulia Ramadhani, Shanita Rambharos, Gaia Redaelli, Jingyuan Ren, Jurgen van der Heijden, and Zihao Zhu. Their cutting-edge research, conducted across diverse geographical contexts from Europe to Asia and the Pacific, Africa and the Americas, offers invaluable insights that advance the discourse on equitable and sustainable cities.

We are also thankful to Prof. Bharat Dahiya, Director of the Research Center for Sustainable Development and Innovation, School of Global Studies, Thammasat University, who also serves as the Series Editor of *Advances in 21st Century Human Settlements,* for his insightful Forewords, which frame this book's discourse for the benefit of the readers.

Finally, the Editors sincerely thank Dr. Loyola D'Silva, Executive Editor, Ms. Shalini Monica Clement Selvam, Ms. Diya Ma, and Ms. Manopriya Saravanan, together with Springer's entire team. Their professional guidance, support and cooperation throughout the preparation and publication process have been instrumental in ensuring high quality and timely release of this volume.

Contents

Editors and Contributors

About the Editors

Mahesti Okitasari is a Research Fellow and Academic Associate at the United Nations University Institute for the Advanced Study of Sustainability (UNU-IAS). Her expertise includes power and politics in multi-level climate and SDG governance, policy and institutional analysis at the interface of climate and development, subnational climate finance, renewable energy diffusion, sustainable consumption and production, sustainable urban development, and SDG localisation at the national and local levels. Her current work covers the interactions between the global agenda and commitments on decarbonisation. She is interested in understanding the role of governance in shaping transformation towards bridging social, economic, and environmental sustainability gaps.

Paul Hudson is a Lecturer in environmental economics at the University of York, who works interdisciplinarity upon natural hazard risk management, Climate Change Adaptation, nature-based solutions, and natural capital. He is currently the Co-chair of the Finance and Economics KAN for Future Earth. Paul works across topics and fields relating to how we can be better prepared for the eventual worsening impacts of climate change and what options are at our disposal for doing so. For example, how can we encourage the use of property level flood protection measures, how effective these measures are, what are the intangible impacts of flooding, and the use of novel insurance mechanisms to support resilience. He is increasingly working on using nature-based solutions to build our stock of natural capital and use the resulting ecosystem services to hit multiple social and environmental goals simultaneously in a manner that is ecological and socially just.

Masachika Suzuki is a Professor at Graduate School of Global Environmental Studies at Sophia University. The area of his research interests includes clean energy technology innovation and development, sustainable finance and banking, and sustainable tourism and community development. He received Ph.D. from Erasmus

Universiteit Rotterdam in the Netherlands and MAs both from Columbia University in the USA and Keio University in Japan. Apart from the academic career, he has work experience in the field of environmental and energy consultancy. The positions include senior analyst positions at Mitsubishi Securities in Tokyo and Innovest Strategic Value Advisors in New York as well as consultant positions at the UN headquarter (UN DESA), the United Nations Framework Convention on Climate Change (UNFCCC), and UNU Institute for the Advanced Study of Sustainability (UNU-IAS).

Contributors

Mohamad Nadim Adi Interior Design, School of Family and Consumer Sciences, Texas State University, San Marcos, TX, USA

Ajibola Anthony Akanji Faculty of Law & Faculty of Management and Social Sciences, Lead City University, Ibadan, Oyo State, Nigeria

Mais M. Aljunaidy Department of Psychology, College of Liberal Arts, Texas State University, San Marcos, TX, USA

Blessings Chiwosi College of Humanities and Development Studies/College of International Development and Global Agriculture, China, Agricultural University, Haidian District, Beijing, China

Catalina Cruz-Piedrahita Higher School of Agriculture and Forestry Engineering, Department of Agriculture Economics, Universidad de Córdoba, Córdoba, Spain

Maria Mar Delgado-Serrano Higher School of Agriculture and Forestry Engineering, Department of Agriculture Economics, Universidad de Córdoba, Córdoba, Spain

Hade Dorst TNO, Vector, The Hague, The Netherlands

Paul Hudson Department of Environment and Geography, University of York, Hesilington, York, UK

Eric Lau Beijing Circular Economy Strategies, Beijing, China

Li Li College of Humanities and Development Studies/College of International Development and Global Agriculture, China, Agricultural University, Haidian District, Beijing, China

Wenjie Liu United Nations University Institute for the Advanced Study of Sustainability, Tokyo, Japan

Maiko Nishi Department of Global Coexistence Studies, College of Bioresource Sciences, Nihon University, Fujisawa, Japan

Mahesti Okitasari United Nations University Institute for the Advanced Study of Sustainability, Tokyo, Japan

Deanty Mulia Ramadhani School of Global Environmental Studies, Sophia University, Tokyo, Japan

Shanita Rambharos TNO, Vector, The Hague, The Netherlands

Gaia Redaelli Escuela Técnica Superior de Arquitectura, Universidad de Sevilla, Sevilla, Spain

Jingyuan Ren Institute of International and Regional Studies, Southeast University, Nanjing, China

Masachika Suzuki Graduate School of Global Environmental Studies, Sophia University, Tokyo, Japan

Jurgen van der Heijden TNO, Vector, The Hague, The Netherlands

Zihao Zhu Global Business Development, TMF Group, Tokyo, Japan

An Introduction to Equitable and Sustainable Cities—Economics and Disruptive Technologies

Paul Hudson, Mahesti Okitasari, and Masachika Suzuki

1 Introduction to Equitable and Sustainable Cities

The Sustainable Development Goals (SDGs), and the Millennium Development Goals before them, highlight that sustainable development is a key global policy goal. This is directly reflected in the United Nations (UN) Sustainable Development Goal 11, which aims to '[m]ake cities and human settlements inclusive, safe, resilient and sustainable' (United Nations 2015). This is an important goal because cities will be at the core of humanity's lived experience. We can highlight this by noting that by 2022 over half of the global human population, over 4 billion, lived in urban areas and this ratio is expected to grow to over 70 per cent by 2050 (United Nations 2023a, b).

However, despite urban residents currently accounting for over 50 per cent of the global population, cities only occupy 3 per cent of the Earth's surface (United Nations 2023a, b). Cities, therefore, are in effect a disproportionate driver of humanity's global environmental, social and economic outcomes. For example, cities may account for 60–80 per cent of energy consumption and 75 per cent of carbon emissions (United Nations 2023a, b). However, these numbers may not be disproportionate when we consider that globally urban areas may account for 80 per cent of global GDP (United Nations, 2023b). Therefore, it is also not surprising that cities are also a locus of increasingly severe impacts from climate change and natural disasters due to their high concentration of people, location and varying capacity to adapt and change to the new environmental and social conditions we face (Dodman et al. 2023).

P. Hudson (✉)
Department of Environment and Geography, University of York, York, UK
e-mail: paul.hudson@york.ac.uk

M. Okitasari
United Nations University Institute for the Advanced Study of Sustainability, Tokyo, Japan

M. Suzuki
Graduate School of Global Environmental Studies, Sophia University, Tokyo, Japan

M. Okitasari et al. (eds.), *Equitable and Sustainable Cities*, Advances in 21st Century Human Settlements, https://doi.org/10.1007/978-981-92-2793-8_1

Considering this potential for rapid and extended urbanisation by the mid-point of the century and how increased urbanisation will drive increasing environmental, social and economic impacts (Ritchie et al. 2024), we must make sure that cities do not develop at a pace that outpaces the development of housing, infrastructure and supporting services that allow cities to be sustainable. Sustainability is a contested topic with multiple different views of the future (Nielsen et al. 2025). However, a common perspective is that of the Brundtland Commission which stated that sustainable development means that we achieve the goals of the present without undermining the ability of future generations to meet their own goals (World Commission on Environment and Development 1987). This is in turn often separated into the pillars of social, economic and environmental outcomes to guide thinking on where potential objectives and the resources to achieve them can come from (Purvis et al. 2019).

Acting upon and across these three pillars has not always been achieved, as we have historically neglected environmental outcomes in favour of economic outcomes leading to large-scale environmental degradation and exceeding several of our planetary boundaries (Richardson et al. 2023). Therefore, to achieve sustainable urban development we may have to significantly transform how our cities are built and managed. For instance, a majority of the projected urban growth could take place in smaller cities in the developing world (IPCC 2023), which stands to outstrip their ability to cope, in turn exacerbating inequalities and urban poverty (Liddle 2017). The number of urban residents currently living in urban slum-like conditions currently stands at over 1 billion and is expected to triple by 2050 (United Nations 2023a).

This example highlights that we cannot address the needs of sustainable development without also considering the power dynamics consistent with how urban areas are used and developed (Mensah 2019). The problems we face as a global society need to be addressed in such a way that they promote rather than undermine long-run sustainable development. For example, development can lead to increased levels of inequality can lead to unrest and insecurity (Murshed 2008), pollution deteriorates health (Liao et al. 2023) and productivity (Wang et al. 2022), while natural disasters directly harm people (IPCC 2023) and an increasing concentration of these threats can undermine our ability to limit them.

Therefore, an equitable and sustainable city is one that balances social justice, environmental responsibility and economic viability while ensuring that all residents, regardless of socioeconomic status, have access to essential resources and opportunities. These cities must be inclusive, resilient and regenerative, addressing both present needs and long-term sustainability goals. However, there is no universal pathway to achieve this goal as how this goal will manifest will strongly depend on what priorities are and why. The studies and perspectives collected in this book explore these pathways either directly or indirectly by exploring how novel or emerging technologies can be used to support, identify, or inhibit action towards a more sustainable future. **Delgado-Serrano, Redaelli and Cruz** in Chap. 2 demonstrate how patios—inner courtyards—have served as a retro-innovative and disruptive technology to avert the globally looming climate crisis by combining plant evapotranspiration, water bodies, architectural features and wind circulation to create passive cooling systems that reduce energy consumption and greenhouse gas emissions.

However, this book also recognises that novel technology itself may not be enough and that not only can novel technologies be used, but so can novel social and economic institutions as well. For example, **Zhu, Ramadhani and Suzuki** in Chap. 3 argue that energy poverty persists even in developed countries despite the deployment of technologies and a comprehensive understanding of the situation, particularly in small and rural communities, is largely missing. A new way of thinking and approaching novel technologies is needed. **Akanji and Hudson** in Chap. 4 consider how housing cooperatives can be used to create more sustainable environments and foster low-carbon building practices by changing the incentive structure individuals face. These housing cooperatives can in turn be supported by novel financial mechanisms and emerging technologies, if the social institution and governance that they support is socially responsible to begin with. **Li, Ren and Chiwosi** in Chap. 5 investigate how communities managed to adopt a collaborative approach to waste management by aligning policy, technology and community agency. Their study shows that by redefining waste management as a collective asset, there is potential for converging novel technologies with institutional innovation. Furthermore, **Nishi** in Chap. 6 explores how to harmonise and effectively pair Indigenous and local knowledge with modern science and technologies for climate-smart agriculture. The chapter specifically highlights how farmers' ecological knowledge and local experiential knowledge can have multiple effects beyond achieving 'triple wins' of productivity, resilience and carbon sequestration (World Bank Group 2012), such as biodiversity conservation and livelihood improvement. These studies further confirm that technology and the social context in which it is developed and used cannot be divorced from one another and we must consider the most socially responsible way of deploying novel technologies so that their ability to disrupt economic and social institutions can be harnessed for the public good.

The role of the public and private sectors should also be explored more in cultivating equitable and sustainable cities and closing financing gaps. **Liu, Okitasari and Lau** in Chap. 7 discuss the strategic role and contributions of green and sustainable municipal bonds in expediting sustainable urban transformation through accelerated infrastructure investment. **Zhu and Suzuki** in Chap. 8 explore the opportunity of introducing solar securitisation—a financial strategy that issues tradable securities backed by future electricity revenue from renewable projects—as a strategy for developing economies that can yield co-benefits of providing energy security and alleviating poverty. The dynamics of novel financing, emerging technologies and co-benefits are important to explore as they define who is gaining, who is bearing the cost, and how they are materialised in the implementation. **Adi and Aljunaidy** in Chap. 9 argue that financial technology has the potential to become a powerful tool with its ability to extend financial services to underserved populations and narrow the financial gap between different socioeconomic groups. They further caution against the risk of the unequal spatial distribution of technological benefits as smaller firms and developing countries may struggle to keep pace with the investment cost of advanced systems like AI, blockchain and quantum computing. While **Hudson** in Chap. 10 considers how similar novel technologies can be used to support insurance as a mechanism for addressing increasing disaster losses in a socially responsible manner.

It is clear that new entrants in emerging financial technology markets have the appetite for non-traditional ways of financing. It is primarily because of difficulties in accurately assessing the value of intangible assets of these young technological firms, limited cash flow and too short a history of performance and profitability (Sahut et al. 2021). It is, therefore, necessary to have appropriate methods to grow novel financial mechanisms, which means an increasing need to overcome barriers and make governance as well as finance fit for such purposes. This endeavour, however, is not devoid of challenges. **Van der Heijden, Rambharos and Dorst** in Chap. 11 found that despite economic and financial appetite in favour of emerging financial strategies, these may not be sufficient to overcome the decision-making hierarchy.

While many of the chapters presented in this book focus on cities directly, others do so indirectly. This is because cities play a critical role in shaping the future of the planet. As hubs of economic activity, population growth, innovation and governance, they are both major contributors to global challenges and essential sites for solutions. Their density, infrastructure and governance structures make them uniquely positioned to drive change in climate action, social justice and economic development. Therefore, even if a chapter does not directly refer to cities and their need to be sustainable and equitable as cities exist at the intersection of multiple social and environmental issues they will become the 'living laboratories' that determine if these social, economic and environmental interventions will be successful or not, and at the scale we need for a more sustainable and equitable future.

References

Dodman D, Hayward B, Pelling M, Castan Broto V, Chow W, Chu E, Dawson R, Khirfan L, McPhearson T, Prakash A, Zheng Y, Ziervogel G (2023) Cities, settlements and key infrastructure. In: Pörtner H-O, Roberts D-C, Tignor M, Poloczanska ES, Mintenbeck K, Alegria A, Craig M, Langsdorf S, Löschke S, Möller V, Okem A, Rama B (eds) Climate change 2022—impacts, adaptation and vulnerability. contribution of working group II to the sixth assessment report of the intergovernmental panel on climate change. Cambridge University Press, pp 907–1040. https://doi.org/10.1017/9781009325844.008

IPCC (2023) Climate change 2021—the physical science basis (V. Masson-Delmotte, P. Zhai, A. Pirani, S. L. Connors, C. Péan, S. Berger, N. Caud, Y. Chen, L. Goldfarb, M. I. Gomis, M. Huang, K. Leitzell, E. Lonnoy, J. B. R. Matthews, T. K. Maycock, T. Waterfield, O. Yelekçi, R. Yu, & B. Zhou, Eds.). Cambridge University Press. https://doi.org/10.1017/9781009157896

Liao L, Du M, Chen Z (2023) Environmental pollution and socioeconomic health inequality: evidence from China. Sustain Cities Soc 95:104579. https://doi.org/10.1016/j.scs.2023.104579

Liddle B (2017) Urbanization and Inequality/Poverty. Urban Sci 1(4):35. https://doi.org/10.3390/urbansci1040035

Mensah J (2019) Sustainable development: meaning, history, principles, pillars, and implications for human action: literature review. Cogent Soc Sci 5(1). https://doi.org/10.1080/23311886.2019.1653531

Murshed SM (2008) Inequality, indivisibility and insecurity. In: Intra-state conflict, governments and security. Routledge, pp 74–91

Nielsen P, Christensen AA, Bolwig S (2025) Comparing contested sustainabilities: how diverse human–nature relationships give rise to different approaches to sustainability. Sustain Sci 20(1):55–76. https://doi.org/10.1007/s11625-024-01595-2

Purvis B, Mao Y, Robinson D (2019) Three pillars of sustainability: in search of conceptual origins. Sustain Sci 14(3):681–695. https://doi.org/10.1007/s11625-018-0627-5

Richardson K, Steffen W, Lucht W, Bendtsen J, Cornell SE, Donges JF, Drüke M, Fetzer I, Bala G, von Bloh W, Feulner G, Fiedler S, Gerten D, Gleeson T, Hofmann M, Huiskamp W, Kummu M, Mohan C, Nogués-Bravo D, Rockström J (2023) Earth beyond six of nine planetary boundaries. Sci Adv 9(37). https://doi.org/10.1126/sciadv.adh2458

Ritchie, H., Samborska, V., & Roser, M. (2024, December). *Urbanization.* Our World in Data.

Sahut J-M, Iandoli L, Teulon F (2021) The age of digital entrepreneurship. Small Bus Econ 56(3):1159–1169. https://doi.org/10.1007/s11187-019-00260-8

United Nations (2015) Transforming our world: the 2030 agenda for sustainable development

United Nations (2023a) SDG 11: sustainable cities and communities. Make cities and human settlements inclusive, safe, resilient and sustainable. In: UN system SDG implementation database. United Nations, pp 1–2

United Nations (2023b) The sustainable development goals report. Special edition

Wang C, Lin Q, Qiu Y (2022) Productivity loss amid invisible pollution. J Environ Econ Manag 112:102638. https://doi.org/10.1016/j.jeem.2022.102638

World Commission on Environment and Development (1987) Our common future

Climate Action, Social Justice, and Nature Preservation

The Role of Patios in Creating Equitable and Sustainable Cities

Maria Mar Delgado-Serrano, Gaia Redaelli, and Catalina Cruz-Piedrahita

1 Introduction

The year 2023 has been officially confirmed as the hottest on record, with unprecedented temperatures largely driven by the accelerating effects of global warming (Wong et al. 2023). As the global population continues to concentrate in urban areas (UN 2015), the accompanying urban development introduces substantial land modifications that profoundly alter the urban microclimate (Loibl et al. 2021). Such modifications exacerbate the urban heat island effect (UHI), further intensifying temperature rises within cities (Phelan et al. 2015) and impacting the health and well-being of urban residents (Pineo and Rydin, 2018). The urgent need to mitigate and adapt to climate change places mounting pressure on cities to re-evaluate and innovate in urban planning and design.

Heat in urban areas, like many other environmental factors, tends to disproportionately affect poorer and more vulnerable populations, such as poor people, children and the elderly (Chakraborty et al. 2019; Green et al. 2019). In large cities, higher population densities are often found in less advantaged neighbourhoods (Burton 2000), which also tend to have fewer green spaces, a pattern linked to the neighbourhood's GDP per capita (Schüle et al. 2017). Hence, local temperatures in these areas can be higher than those experienced across the city as a whole (Benz et al. 2021). For instance, homes in wealthier neighbourhoods often have private green spaces and outdoor recreational areas, whereas such features are typically absent in less advantaged areas.

M. M. Delgado-Serrano (✉) · C. Cruz-Piedrahita
Higher School of Agriculture and Forestry Engineering, Department of Agriculture Economics, Universidad de Córdoba, Córdoba, Spain
e-mail: mmdelgado@uco.es

G. Redaelli
Escuela Técnica Superior de Arquitectura, Universidad de Sevilla, Sevilla, Spain

M. Okitasari et al. (eds.), *Equitable and Sustainable Cities*, Advances in 21st Century Human Settlements, https://doi.org/10.1007/978-981-92-2793-8_2

Air conditioning, while effective at cooling indoor spaces, contributes to heating the surrounding environment and requires significant energy consumption, making it an inequitable and environmentally unsustainable solution (Randazzo et al. 2020). Currently, urban development in warmer regions often relies on technological strategies that exacerbate inequities. People with fewer financial resources are less able to afford air conditioning during extreme heat events (Hayden et al. 2017), when demand typically drives temperatures even higher, or to purchase specialised materials, such as light-filtering glass. Moreover, these individuals often lack access to nearby green spaces for cooling, as such spaces are generally absent in disadvantaged areas (Chakraborty et al. 2019; Green et al. 2019).

When these economic challenges are worsened by additional barriers, such as limited mobility among elderly individuals, mothers with infants, or those without private vehicles to reach cooler areas (Dumlao et al. 2023), vulnerable populations are effectively confined to living in environmentally detrimental conditions. These conditions can have severe health consequences, even leading to fatalities (Rohat et al. 2019). Prolonged exposure to high temperatures can lead to several mental and health issues (Ebi et al. 2021; Liu et al. 2021; Singh et al. 2024). Moreover, people with chronic diseases are more prone to die due to heat-related problems (Singh et al. 2024). A recent study estimated that in 2022, approximately 61,672 heat-related deaths occurred across Europe (Ballester et al. 2023).

Furthermore, other environmental and social issues that are already problematic in urban areas worsen, with high temperatures, air pollution (Stone 2005) and social isolation (Kim et al. 2020) being two of those, and they are increasingly recognised as threats to mental health (Bakolis et al. 2021; Tung et al. 2019). Exposure to air pollution, particularly pollutants like particulate matter (PM2.5) and nitrogen dioxide, has been linked to heightened risks of anxiety, depression and even cognitive decline due to its neuroinflammatory effects (Buoli et al. 2018), and normally, the poorer neighbourhoods tend to be more prone to bad quality air due to their closeness to roads, industry and their lack of green space (König 2024). Additionally, under heat stress conditions, people tend to stay in close environments, avoiding high temperatures, leaving vulnerable populations, such as the elderly and those with pre-existing mental health conditions, at a higher risk of isolation (Kim et al. 2020). Social isolation, combined with urban environments lacking accessible green and communal spaces, can lead to feelings of loneliness and social anxiety (Turan and True 2023). This isolation can deepen the psychological impacts of environmental stressors, creating a feedback loop that worsens mental health outcomes. High temperatures, with other environmental and social conditions, make the current urban environment in cities in hot and arid climates a highly inequitable and unsustainable one. Cities are demanded to create healthier and more resilient environments for citizens while identifying cost-effective strategies to enhance urban liveability and overall well-being. This challenge is critical as cities seek sustainable solutions to balance rapid urbanisation with the imperative to reduce environmental impact.

In the Mediterranean cities, where summer temperatures are already extremely high, the future temperature landscape is expected to experience further impacts due to challenges from climate change (Paz et al. 2016). Traditional architecture in these cities often incorporates patios, the Spanish name for inner courtyards, that can be an economical and retro-innovative technology for climate change adaptation (Callejas et al. 2020). Patios are open spaces within the interior of buildings, typically with plants and small bodies of water (Petruccioli 2004) and, historically, patios as multifamily houses in the most marginal neighbourhoods have been communal spaces in residential settings, housing kitchens and laundry areas, where residents could cook, clean and socialise with neighbours, which makes them an inclusive type of green space. More recently, these spaces also serve as gathering places, often with adjacent dining or living areas, where families and friends meet to eat, play and engage in social activities (de Canales et al. 2012).

This chapter's objective is to describe how courtyards might contribute to temperature reduction and enhance environmental conditions for individuals in their vicinity without necessitating substantial financial investment, as is often required by other technological solutions. The combination of vegetation and architectural elements in patios serves as a passive cooling system, offering shade, promoting ventilation and enhancing outdoor living spaces (Taleghani et al. 2014) without the energy demands of other cooling systems like fans or air conditioning. Integrating well-designed patios into architectural designs represents an effective, affordable and aesthetically pleasing strategy to reduce reliance on active cooling systems. This leads to energy savings and a more sustainable built environment (Zamani et al. 2018), especially for those with lower living standards who cannot afford the high energy prices. Furthermore, the chapter examines public perceptions of courtyards to assess whether they are regarded as a user-friendly solution for urban residents rather than an imposed measure, probably serving as spaces for socialisation, community building and quality of life (Bassal et al. 2023). The overall aim is to propose an alternative that is economically and socially equitable, as well as environmentally and socially sustainable. By considering the multifunctional roles of patios, both as spaces for social interaction, especially for people in poor neighbourhoods, and as contributors to ecological sustainability, the chapter argues that they represent an economical, practical, yet highly useful tool in creating more equitable, resilient and sustainable cities.

2 Methods

This research is conducted in IN-HABIT (https://www.inhabit-h2020.eu), a European-funded H2020 project investigating how to boost inclusive health and well-being in four European cities: Cordoba (Spain), Lucca (Italy), Riga (Latvia) and Nitra (Slovakia). In Cordoba, IN-HABIT investigates the role that patios can play in facing the challenges of climate change, focusing on vulnerable collectives. The project assesses using a mixed-methods approach to the social and ecological comfort conditions in the patios of the historical centre and provides information on how the characteristics in terms of vegetation, architecture and soil permeability influence their performance as passive cooling systems. It also analyses the perception of patios as places of socialisation that boost the health and well-being of urban dwellers. Sensors were strategically placed to collect data in patios of different characteristics, and people were interviewed about the importance of patios in the city, enabling the analysis of the well-being levels of inhabitants, both physical and perceived, correlated with the patio type and vegetation. This data, combined with insights from interviews, aims to provide a comprehensive understanding of how the ecology and anthropology of the Mediterranean patios contribute to better thermal comfort, community building and affordable cooling systems.

2.1 Córdoba Case Study

Cordoba is a medium-sized city (325,916 inhabitants) located in Andalusia, South Spain. It is characterised by a 'subcontinental Mediterranean climate with very hot summers'. It is also one of the hottest cities in Spain (AEMET 2024) and one that is suffering the most from climate change. Global circulation models for the climate change projection in Andalusia estimate that the average temperature will increase between 2 and 4ºC and that most of this increase will fall in the spring and summer. Studies predict that the cooling rate in regions such as the capital of Cordoba will increase at the end of this century between 39.7 per cent in the best scenario (CNCM3 B1) and 85.8 per cent in the worst (Consejería de Sostenibilidad y Medio Ambiente 2022). This increase in energy needs for air conditioning in the context of rising energy prices is extremely critical, with the consequent risk of pushing many households into energy poverty and reducing the competitiveness of companies. Many other areas, such as mobility and the strategic tourism sector, may be delicately affected (Gabaldón-Leal et al. 2015).

The city has a long history and cultural background dating back to Roman times. It was the city capital of the Al-Andalus Arab Empire (eighth–thirteenth centuries). These centuries' wealth left a unique cultural heritage; Cordoba is the first city to hold four inscriptions on the UNESCO World Heritage List. Part of the city's intangible heritage is the presence of patios, which UNESCO recognise for their urban, architectural and anthropological value. Patios are a clearly distinctive city element from ancient times to the point that up to this date, one of the most visited events is the '*Fiesta de los patios*' where people open their patios for others to visit (González et al. 2019). The casas-patio is a courtyard house inhabited by one or several families, as in the *casa de vecinos (neighbourhood house),* that was first implemented in the Domus in Roman time but had an important development in the Islamic city, as in the case of Córdoba. As a meeting place and as an interior garden, the patio houses occupy part of the urban fabric of Córdoba and enclose not only urban and architectonical values but also ecological aspects that guarantee the patio and its sociality. However, today, most of those traditional uses have been left behind and are being replaced with tourist activities or just as vacant land (Redaelli 2019).

The case study of Córdoba is emblematic because of the dimension of the historical centre (i), its connection with the environment, thanks to the presence of the Sierra

Table 1 List of patios and their characteristics

	Size (Square meters)	Shape	Type of soil	Vegetation coverage
Patio 1	6	Square	Impermeable like asphalt	Low
Patio 2	142	L	Permeable surface	High
Patio 3	30	Square	Permeable surface	Moderate
Patio 4	77	Square	Permeable surface	Moderate

Morena and the Guadalquivir River (ii) for the urban and architectonical structure of the urban heritage within the courtyards (iii) and for the cultural link the city has with patios (iv). Hence, the city (especially the historical centre) works as a model of how patios work as an eco-social adaptation to temperature that creates sustainability and equity benefits that can possibly be extrapolated to other Mediterranean cities.

2.2 *Quantitative Analysis of Temperature Inside Patios*

To understand how courtyard characteristics influence temperature, sensors were installed in four pilot patios in the historical centre of Córdoba, selected based on different values, among them, (i) the dimension and geometry of the void, (ii) the degree of permeability of the soil, (iii) the presence of rooted and non-rooted vegetation and (iv) the height versus wide ratio. In each patio, one sensor was positioned at a higher level (approximately 400 cm) and another at a lower level (approximately 150 cm). The sensors collected data from July 2023 to August 2024, with temperature readings recorded every fifteen minutes on average. However, for the purposes of this study, only the maximum, minimum and average temperatures were utilised. This chapter presents the results of four patios with different characteristics (Table 1).

2.3 *Qualitative Analysis of People's Perception of Patios in Córdoba*

To understand the social and ecological role of courtyards in the city of Córdoba, we interviewed nine people. Three were experts in green spaces or patios, an architect working with courtyards, a botanist specialising in urban green spaces, and a resident of Córdoba's historic centre who has preserved the patio tradition for over 30 years, three were residents from the city centre randomly selected, and three were residents from an underserved neighbourhood lacking green spaces, where courtyards consist of large concrete areas between buildings. The latter served as a control to determine if the concept of the patio was ingrained in the population and whether their opinions differed from those with access to courtyards. We asked all participants four questions: 1) What is a green space? 2) What is the importance of green space? 3) What is a patio? 4) What is the importance of a patio? Their responses were recorded on video and later transcribed.

3 Results

We present the results from our literature analysis and those from our empirical research. The results reveal that courtyards function as both social and ecological hubs, highlighting their potential to promote more sustainable urban environments. The literature review identified how patios have been used over the centuries in hot areas worldwide, their ecological benefits, the correlation between courtyards and well-being or health, and the characteristics that permit them to work as passive cooling systems. The empirical research using sensor data and interviews validated the influence of patio features in its functioning and the role of patios as social places. We divided this section into the ecological benefits of patios by showing the information related to the topic we found in the literature and the results of the sensors' analysis. Then, we show the socio-cultural part of patios, including the interview results and the information found in the literature.

3.1 Patios as Green Spaces and Passive Cooling Mechanisms

Patios have been integral to architecture for centuries, particularly in regions with hot climates, due to their ability to function as passive cooling systems. Traditional houses were articulated around an inner patio with water (well, fountain or pool) in the centre and plants that generate freshness, providing climatic quality to the built environment. Some studies have recently researched how the combination of natural ventilation, shading and thermal mass creates cooler microclimates, reducing the need for mechanical cooling (Diz-Mellado et al. 2023). Shadow strategy, both for the architectonical composition thanks to the patio and elements like '*toldos*', traditional awnings used in the summertime during the daytime, are relevant methods to produce better thermal conditions. Nevertheless, vegetation is one of the most relevant elements that create better conditions for a more liveable environment and reduce energy consumption (Sun et al. 2021).

The disruptive, equitable and sustainable innovation we propose is to use patios to address climate change. The system is based on nature, plant evapotranspiration and other characteristics as follows. Evapotranspiration is a key cooling process that plants use to regulate temperature, making it one of the most significant energy exchange mechanisms on a global scale (Cascone et al. 2019). Through transpiration, plants extract approximately 585 kcal of energy from the atmosphere for each litre of water they release via their leaves at a temperature of 20ºC, effectively cooling their surroundings. Plant transpiration can dissipate between 30 per cent and 70 per cent of the sun's energy as latent vaporisation heat, reducing temperatures by as much as

10°C on hot, dry days (Guerrero et al. 2016). The process allows the excess energy to be dissipated at a rate equivalent to the latent heat of vaporisation. However, if the patio is significantly wider than its height, and a large proportion of its surface is exposed to direct sunlight throughout the day, the cooling effect becomes insufficient. In such cases, the patio may, in fact, be warmer than the rest of the house. Therefore, it is crucial to consider the appropriate amount of greenery, the materials used and the overall geometry of the structure to optimise its thermal performance.

Furthermore, according to Guerrero et al. (2016), the cooling effect of evapotranspiration doesn't require water to reach a boiling point. Instead, it occurs as water on plant surfaces meets dry air, triggering evaporation. Unlike passive cooling through shadows provided by structures, plant transpiration uniquely converts a substantial amount of solar energy into latent heat, cooling the environment as water vapour is produced. In thermodynamic terms, this effect is known as 'adiabatic cooling', with only ammonia surpassing water's cooling efficiency. This cooling capability depends on factors like ambient temperature, wind speed and relative humidity. For instance, in a controlled environment with a temperature of 35°C and 18 per cent relative humidity, water evaporation can cool the space to 19.2°C, achieving 90 per cent relative humidity. As an example, a tree with a 50 m2 crown in a healthy garden in Cordoba, with available water, can have a cooling capacity of 11.3 kW in the month of August. This power is equivalent to 8 air conditioners of 1.5 kW of electrical power serving 400 m2 of an enclosed space. However, for this cooling process to be effective, plants must have an adequate water supply, and the efficiency is heightened in dry atmospheric conditions. Without water, evapotranspiration significantly decreases or ceases entirely.

In open areas, the cooling effect of evapotranspiration becomes perceptible only when there is a sufficient volume of plants and water (Ghafarian et al. 2022). A critical mass of well-irrigated vegetation is essential to create a microclimate that effectively cools the surroundings. On the contrary, houses with patios offer small-scale examples of this passive temperature regulation. Even with a relatively small critical mass of vegetation, the enclosed walls of a patio prevent the cooling effect from dispersing, allowing it to function independently with less influence from external climate conditions. However, it is important to consider that several factors can enhance this cooling effect. Permeable pavements that allow rainwater infiltration, well-sized trees and carefully managed irrigation all play a role. Irrigation control is essential, not just in the amount of water used but also in the timing and frequency. For example, watering at dusk may be less efficient in terms of water conservation but quickly lowers the ambient temperature, producing a more immediate cooling effect in the area. Water bodies in the patios are also very important not only because of evaporation and direct cooling from the water sheet but also because of their high thermal inertia.

A high and diverse canopy makes evapotranspiration in patios more effective. Trees play a critical role in this operation by releasing moisture into the air through transpiration, which cools the surrounding air. The cooler air then descends to ground level, providing a more comfortable microclimate below. This can be achieved by creating a structured, layered arrangement of various trees and plants that extends continuously from ground level upward. For optimal results, the plant species selected should be well-adapted to hot, dry climates. These plants not only can conserve water effectively but also have a high transpiration efficiency, meaning they can continue releasing moisture without stress when well-supplied with water and can shut down their transpiration (by closing their stomata) to survive drought conditions or extreme heat. For the case of Cordoba patios, Mediterranean plants are particularly well-suited for this purpose, as they naturally balance these characteristics.

Air circulation also has an important role in cooling systems at two levels. One is linked to the architectural design and the possibility of wind circulation, but, more importantly, the other is linked to convection (vertical air movement) and advection (horizontal air movement) processes in plants (Salisbury and Ross 1992). Wind may have a greater influence on heat removal by air circulation than by vapour withdrawal. Advection removes the vapour resulting from transpiration, reducing the thickness of the boundary or stationary layer of the leaf, which substantially increases water loss, especially if it is dry. Secondly, advection is responsible for the removal (or contribution) of heat by convection, which plays a significant role in enhancing the cooling effects of evapotranspiration, particularly during summer months when relative humidity is low. Under these dry conditions, the wind can have a drying effect, increasing the rate at which plants transpire and release moisture. In a well-watered patio, this airflow amplifies the heat removal process in the upper parts of the canopy. At the same time, cooler air remains trapped within the lower, protected layers of the foliage, creating a refreshing atmosphere. However, this cooling effect reverses when air temperatures become extremely high. At very high temperatures, advection can start to introduce heat instead of removing it. When temperatures approach 35ºC, the benefits of advection in dispersing heat begin to diminish, and when temperatures exceed 40ºC, the wind's effect can actually be counterproductive, adding heat to the environment rather than cooling it. This makes atmospheric stability, where there's minimal wind, a considerable advantage in the case of Córdoba, as it helps retain the cool, moist air in shaded areas, sustaining a pleasant microclimate even during extreme heat (Guerrero et al. 2016).

The microclimatic performance of patios is highly influenced by their size, shape, materials and design. Studies indicate that a courtyard's aspect ratio (AR), the width-to-length ratio, affects its ability to regulate temperatures, with an AR between 1 and 2 providing optimal shading in Mediterranean climates (Rojas-Fernández et al. 2017). The height of surrounding structures contributes to thermal comfort by reducing exposure to direct sunlight and maintaining cooler conditions in the summer months (Martinelli and Matzarakis 2017). Moreover, patios help shield buildings from pollutants and stabilise temperature fluctuations, offering cooler and more comfortable environments than surrounding urban areas. By incorporating vegetation, courtyards also filter pollutants and absorb carbon dioxide, improving air quality (Mexia et al. 2018). These findings highlight the relevance of patios in enhancing urban resilience and liveability, particularly in high-density areas where green spaces are scarce and temperatures are rising (Redaelli 2020).

3.2 Cooling Performance of Cordoba's Patios

In Patio 1, the upper sensor consistently recorded higher temperatures than the lower sensor throughout the year, with differences at times reaching close to 9°. This pronounced trend is logical, as the patio is narrow and tall, meaning that sunlight does not reach the lower part, which tends to be dark and cool. Patio 2 also showed a difference between the upper and lower sensors, though these differences were smaller than in the first patio, averaging between one and four degrees. In this larger patio, sunlight reaches the ground level for much of the day; however, factors such as the permeable ground, L-shaped layout and abundant vegetation contribute to the reduction in temperature (Fig. 1).

The last two patios, of medium size, with semi-permeable ground and moderate vegetation cover, generally showed a difference of 0° to 2° between the upper and lower sensors. However, this difference increased during the summer months, as the semi-permeable surfaces and the presence of plants prevented a large temperature increase, such as the one experienced in the upper area. Temperatures across all patios showed the greatest differences during months with extremely high or low temperatures. In Patios 2, 3 and 4, where there is vegetation, the temperatures at the lower level tend to be more stable. In contrast, in Patio 1, which lacks vegetation cover, temperatures at the lower level drop considerably during the winter months (Fig. 1). This demonstrates that the presence of plants not only helps to prevent temperature increases during hot periods but also contributes to thermal stability during colder months.

3.3 Socio-Cultural Role of Patios

Patios have held considerable social, cultural and historical value across Mediterranean civilisations, particularly within hot and arid regions, where they function as communal spaces and environmental adaptations. Integrated into residential or public architecture, patios provide more than simple functionality; they foster community ties, offer aesthetic pleasure and promote relationships with nature. Over centuries, these spaces have evolved to support social interaction, reflect cultural values and facilitate architectural innovation, becoming resilient and adaptable features of the built environment.

Historically, patios have served as the heart of family homes and neighbourhoods, offering residents a shared area for gathering, relaxing and connecting. Many traditional patio layouts reflect the desire for a private, open-air space where family members can come together, shielded from public view but open to nature (Anaya-Sahuco 2019). Across Mediterranean cultures, these enclosed spaces encourage a sense of belonging and community, reinforcing family ties and neighbourly bonds. In Islamic homes, for example, patios create intimate spaces where families can enjoy the outdoors without exposure to the public eye, fulfilling privacy needs central to

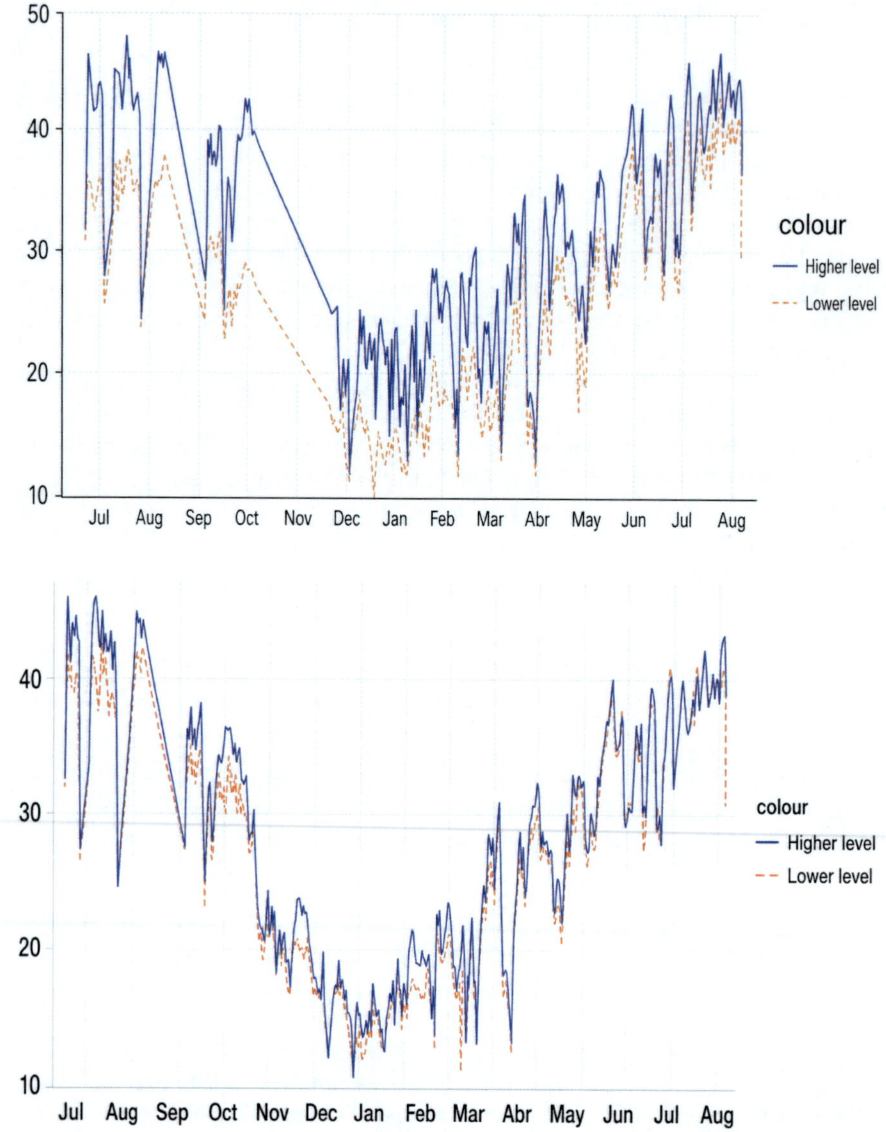

Fig. 1 Seasonal variation of temperatures in Celsius degrees in four patios in Córdoba, Spain. The blue line represents the temperature of the top sensor, and the dotted orange line represents the temperature of the bottom sensor. The temperature sensor in the first patio had malfunctioned between October and December; those differences were not considered

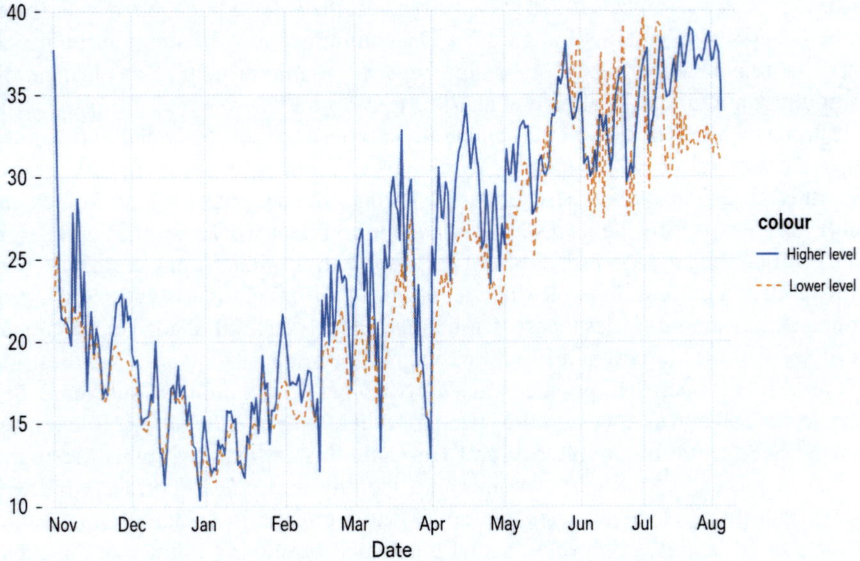

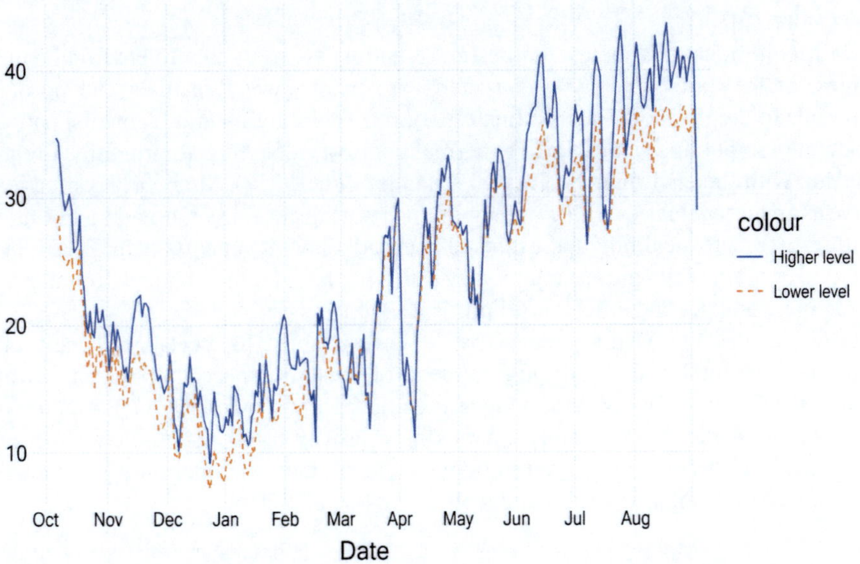

Fig. 1 (continued)

cultural practices (Raviz et al. 2015). In mosques, the *sahn*, a large, open space often embellished with fountains and gardens, is a communal area for ablution rituals and overflow prayer gatherings. This design reflects Islamic principles of purification, community and unity (Frishman et al. 2002).

European Mediterranean cities illustrate another facet of the social and cultural importance of patios, where these spaces have been central to urban life for centuries (Saura et al. 2017). In southern Spain, Italy and Greece, patios within homes and public spaces offer residents a semi-private environment where families and neighbours can gather and socialise, balancing privacy with community engagement. These patios reflect a cultural lifestyle that values both familial closeness and social cohesion, creating shared spaces that foster interaction in a peaceful, shaded environment. Córdoba provides a particularly vibrant example through its annual Patio Festival, where residents open their private patios for two weeks to the public to showcase their lush vegetation and as a welcoming space for others to visit and interact. Recognised by UNESCO as an Intangible Cultural Heritage, this festival highlights the significance of patios as symbols of communal identity and cultural pride (UNESCO 2003). During the festival, residents and visitors engage in a shared appreciation of the patios' beauty, reinforcing social bonds and promoting a sense of collective heritage deeply embedded in the region's architectural traditions (González-Santa Cruz et al. 2019).

In Mediterranean climates, the design of patios is closely tied to environmental values, as they are typically constructed with plants, water features and porous materials to create cooler microclimates. By incorporating natural elements, patios moderate temperature and improve thermal comfort, reflecting a harmonious relationship with the environment (Rojas-Fernández et al. 2017). Mediterranean patios serve as green sanctuaries within homes, often featuring citrus trees, flowers and water features to create a calming atmosphere. The riad's layout supports natural ventilation and cooling while preserving privacy, fulfilling both cultural and environmental needs in a compact and efficient design.

Thus, across the Mediterranean, patios have adapted to meet social, cultural and environmental needs, offering residents a versatile space that reflects local values and lifestyles. Whether serving as communal areas, symbols of cultural identity, or sustainable design features, patios have proven to be essential elements of urban life that strengthen community bonds and enhance the quality of residential environments, promoting health and well-being for inhabitants.

4 Córdoba's People's Perception of Patios

The responses to the first question about what participants considered green space in cities were largely consistent. Most associate green spaces with public areas, such as parks or gardens, defining them as places with plants and flowers. Additionally, many participants noted that green spaces tend to have cooler ambient temperatures, indicating an instinctive association between green spaces, plant diversity and lower temperatures.

4.1 Neighbour from the City: 'It is a Place with Vegetation, for Example, a Park'

In response to the second question on the importance of green spaces, participants described them as recreational areas where people could enjoy themselves and socialise. They emphasised the importance of being in contact with nature and spending time in such spaces. Many highlighted the cooling effect of green spaces, with one participant (a resident from the vulnerable neighbourhood) mentioning the value of being able to smell the flowers, while another (a resident from the city centre) noted the need to increase green space coverage in the city, observing that there was currently an insufficient amount. The similarity in responses across both experts and residents regarding the definition and importance of green spaces suggests that people recognise the significance of green spaces and generally desire close access to them. All participants associated green spaces with improved environmental conditions and emphasised the importance of social interactions, indicating a natural affinity for these spaces within the community.

4.2 Neighbour from Vulnerable Neighbourhood: 'It is a Place Where You Can Go and Smell the Flowers'

However, responses diverged between experts and residents regarding the nature and importance of patios. Experts clearly defined patios as open spaces within a house or building, typically featuring decorative vegetation and traditionally incorporating porous ground or potted plants. They explained that patios are cooler areas within urban environments due to porous ground surfaces and plant evapotranspiration. Experts also highlighted the cultural role of patios in Córdoba, noting their historical use as places for socialisation and community activities, where neighbours would gather for tasks, such as cooking or washing, as water wells were typically located in these communal spaces.

In contrast, neighbours from the city emphasised the importance of patios primarily for social and family gatherings, where friends would meet for activities like sharing a meal over the weekend. These residents highlighted the aesthetic value of plants in courtyards but did not mention any cooling mechanisms or perceive courtyards as particularly cool spaces. Moreover, neighbours from the vulnerable neighbourhood considered any open area near buildings or between buildings a patio. They did not associate patios with vegetation or the other bio-physical aspects identified by the experts. This may be because, in this area, what is referred to as 'patios' are large concrete zones separating building blocks, which tend to be very warm due to heat reflection from the concrete surfaces, and the three neighbours expressed the need to increase the vegetation in their patios and complain that to go to a green area they had to leave the neighbourhood. Finally, all neighbours (both those from the vulnerable neighbourhood and other parts of the city) mentioned the social value of these spaces as places for gathering and interaction, and in the case of the vulnerable neighbourhood, it is also considered a place to go after sunset where temperatures are milder to escape the heat in their houses, because they often experienced energy outages in summer.

5 Discussion: The Role of Patios in Equitable and Sustainable Cities

In this chapter, we examine patios as a sustainable and equitable nature-based solution for, on the one hand, urban heat management, leveraging architectural and ecological elements to cool down temperatures and, on the other hand, increase equity opportunities by creating semi-public/semi-private spaces where people can coexist, especially in vulnerable places. This discussion integrates findings from the literature on patios with our empirical results, gathered through temperature sensors in different patios and interviews with residents from various socio-economic backgrounds. The purpose is to understand how patios function as passive cooling systems and socio-cultural assets in urban environments.

Our results confirm the role of patios as effective passive cooling mechanisms, in alignment with existing literature that highlights the climatic benefits of patios in hot regions. Patios combine natural ventilation, shading and plant evapotranspiration to reduce temperatures. The literature establishes that patios are particularly effective in hot climates due to their enclosed design, which helps retain cool air and minimise the dispersal of heat-mitigating effects (Taleghani et al. 2014). Our sensor data corroborates these findings, showing that patios with permeable surfaces, ample vegetation and shaded areas (Patios 2, 3 and 4) maintained lower and more stable temperatures at ground level compared to the upper levels, especially during the warmer months. This temperature differential aligns with studies like Rojas-Fernández et al. (2017), which notes that an optimal aspect ratio and design features like vegetation coverage significantly enhance a patio's cooling efficacy.

Evapotranspiration, a process identified in previous research as a key cooling mechanism, plays a central role in the patios we studied. As the literature shows, plants release water vapour, which absorbs heat from the surrounding air, reducing the ambient temperature by up to 10°C under ideal conditions (Mexia et al. 2018). In our study, Patios 2 and 3, which had high vegetation cover and permeable surfaces, exhibited lower temperature increases, indicating that plant presence is indeed crucial in mitigating temperature extremes. Patio 1, which lacked vegetation, showed higher temperature fluctuations, particularly during colder months, confirming that the presence of plants not only reduces heat in summer but also stabilises temperatures throughout the year. These findings reinforce the argument that well-irrigated, plant-rich patios contribute to thermal stability and can be used as natural cooling systems in urban design (Soflaei et al. 2016).

Additionally, our study illustrates the importance of architectural factors in enhancing the cooling effects of patios. The L-shaped design of Patio 2, coupled with permeable surfaces, contributed to a temperature difference of 1° to 4° between higher and lower sensor levels, supporting the assertion by Martinelli et al. (2017) that layout and materials influence thermal comfort. The findings in Patio 1, where the narrow and tall structure limited sunlight at the lower level, also highlight the impact of patio dimensions on temperature variation. This observation corresponds with studies on urban microclimates that suggest shading and structural proportions play a substantial role in moderating heat within enclosed spaces.

In terms of socio-cultural relevance, our findings reveal that patios are not only functional but also deeply embedded in the cultural fabric of Mediterranean societies. The literature underscores the social and aesthetic value of patios, where they serve as spaces for family gatherings and community activities (Bassal et al. 2023; Lee and Park 2015). The interviews conducted in Córdoba reaffirm this socio-cultural dimension, with residents from the city centre viewing patios as essential communal spaces, particularly for family meals and gatherings. The annual Patio Festival in Córdoba, as noted by Lopez-Guzman et al. (2016), further underscores the cultural pride associated with patios, illustrating how these spaces serve as symbols of local identity and heritage.

Our results also highlight a socio-economic dimension in how patios are perceived and utilised. While residents in more affluent areas view patios as aesthetically pleasing spaces with greenery that enhances social interaction, residents from vulnerable neighbourhoods often lack access to such well-designed patios, but they do express their wish of having more trees and green elements in their patios, and they use them as heat scape areas during energy outage. For these residents, 'patios' often refer to large concrete spaces between buildings, with limited or no vegetation, which lack the cooling effects found in the greener patios of central Córdoba, but still, they use them often. This disparity suggests that access to open spaces is a natural need of people living in cities, especially under heat stress, but green, well-maintained areas, even patios, may be linked to socio-economic status, highlighting an area where urban planning could address inequity in access to beneficial green spaces. Safarzadeh et al. (2005) suggest that green spaces improve community well-being, and our findings align with this, showing that well-designed patios contribute to physical comfort and social cohesion, which may be less accessible to residents in less affluent areas.

In urban design and planning, patios can serve as an alternative form of green space that may enhance equity and sustainability while reducing energy demand more efficiently than some modern technologies. For instance, Soflaei et al. (2016) noted that patios in historic urban centres are particularly valuable as they allow the incorporation of green infrastructure without compromising architectural heritage.

In densely populated areas, such as city centres or disadvantaged neighbourhoods like those discussed here, large green spaces such as parks may not be feasible due to limited land availability. Moreover, modern green infrastructure solutions like green walls and green roofs are often expensive, hence their implementation in certain urban areas with hot climates could inadvertently exacerbate social inequities, contributing to disparities in health and well-being.

Patios, by contrast, are smaller in scale compared to parks and can be more easily integrated into the urban fabric. They are typically less expensive than other forms of green infrastructure and are widely regarded as socially acceptable. Furthermore, their semi-private or semi-public nature can provide a safer environment than large parks, which may pose safety concerns, particularly for the elderly, women and children. In cultural contexts where traditional norms limit women's access to public spaces, such as in some Arab countries, which often experience extremely hot climates, patios offer an accessible and culturally appropriate alternative.

For individuals with limited mobility, such as older adults, mothers with young children, or others facing similar challenges, patios represent a practical option as they are often directly attached to buildings, ensuring ease of access. While a single patio may not significantly improve environmental conditions over a large area, networks of interconnected patios in adjacent buildings can create green corridors, offering the potential to enhance environmental conditions at a broader scale.

Thus, patios represent an innovative and adaptable solution for urban areas, taking advantage of the interplay of nature and architecture, as well as biotic and abiotic elements, to mitigate heat and promote health and well-being. This approach offers the possibility of developing more equitable and sustainable cities without requiring substantial economic investments, as is often the case with many modern technological solutions.

6 Conclusions

Patios are an element in the creation of equitable and sustainable cities. They represent a multifunctional solution to the challenges of urban living, blending environmental, social and cultural benefits. By serving as natural cooling systems, promoting social cohesion and contributing to environmental sustainability, patios can be crucial to boosting urban resilience. Patios function as passive cooling systems, better than the current technology used like air conditioning, provide green spaces and offer a range of ecosystem services, making them indispensable in the context of urban growth and climate change. Patios contribute to reducing urban heat, lowering energy consumption, enhancing the resilience of cities to environmental pressures and promoting social equity by providing accessible green spaces.

As cities continue to face the challenges of population growth, environmental degradation and social inequality, patios offer a versatile and scalable solution that addresses these issues holistically. By integrating patios into urban planning, cities can improve the quality of life for their residents, reduce their environmental impact and create more sustainable and resilient urban environments. The multifunctional nature of patios, combining environmental, social and cultural benefits, underscores their importance in the future of urban development. Their adaptability to diverse urban contexts and socio-economic settings, along with their cultural significance, permits their replicability in many contexts. As we look towards creating cities that are both liveable and sustainable, patios might play an increasingly important role in shaping the urban landscapes of future equitable and sustainable cities.

7 Declaration of Competing Interests

This research has been funded by the IN-HABIT project (Inclusive Health and Well-being in Medium and Small-Sized Cities—https://www.inhabit-h2020.eu (accessed on 7 November 2024) as part of the Horizon 2020 Programme (grant agreement no.: 869227). The authors thank Jacqueline Ghysbrecht for her support in the collection and curation of data and to the members of PAX: Jacinta Ortiz, Carlos Anaya, Elsa Franceschetto, Paula León and Ángel Molina. Research members of the University of Córdoba: Jose Luís Quero; Marta Baena; and Francisco Javier Martínez. Finally, the authors thank the owners of the patios for allowing entering their homes to install the sensors.

References

AEMET (2024) Resúmenes climatológicos. España. Madrid. Retrieved from https://www.aemet.es/es/serviciosclimaticos/vigilancia_clima/resumenes?w=0&datos=2

Anaya-Sahuco C (2019) Esto no es un patio. Un acercamiento a la complejidad de y en los patios de Córdoba (Universidad Nacional de Educación a Distancia (España). Facultad de Filosofía). Universidad Nacional de Educación a Distancia (España). Facultad de Filosofía. Retrieved from https://hdl.handle.net/20.500.14468/15186

Bakolis I, Hammoud R, Stewart R, Beevers S, Dajnak D, MacCrimmon S, Fecht D et al (2021) Mental health consequences of urban air pollution: prospective population-based longitudinal survey. Soc Psychiatry Psychiatr Epidemiol 56:1587–1599

Ballester J, Quijal-Zamorano M, Méndez Turrubiates RF, Pegenaute F, Herrmann FR, Robine JM, Achebak H et al (2023) Heat-related mortality in Europe during the summer of 2022. Nat Med 29(7):1857–1866

Bassal C, Rabea M, Felix M (2023) A comparative study of mediterranean courtyard houses and the bioclimate impact on their design from four axes: historical, environmental, social and geometry. In: Green building & construction economics, pp 123–137

Benz SA, Burney JA (2021) Widespread race and class disparities in surface urban heat extremes across the United States. Earth's Future 9(7):e2021EF002016

Buoli M, Grassi S, Caldiroli A, Carnevali GS, Mucci F, Iodice S, Bollati V et al (2018) Is there a link between air pollution and mental disorders? Environ Int 118:154–168

Burton E (2000) The compact city: just or just compact? A preliminary analysis. Urban Stud 37(11):1969–2006

Callejas I, Galán-Marín C, Cleonice L, Durante L, Do Amarante E, Guarda M, Rivera-Gómez C (2020) Contrasting the passive cooling effect produced by courtyards located in the Tropical and Mediterranean climates. In: Planning post carbon cities: 35th PLEA conference on passive and low energy architecture, A Coruña, 1st-3rd September 2020: Proceedings, Universidade da Coruña, pp 1686–1691

Cascone S, Coma J, Gagliano A, Pérez G (2019) The evapotranspiration process in green roofs: a review. Build Environ 147:337–355

Chakraborty T, Hsu A, Manya D, Sheriff G (2019) Disproportionately higher exposure to urban heat in lower-income neighborhoods: a multi-city perspective. Environ Res Lett 14(10):105003

Chaparro MP, Benzeval M, Richardson E, Mitchell R (2018) Neighborhood deprivation and biomarkers of health in Britain: the mediating role of the physical environment. BMC Public Health 18:1–13

Consejería de Sostenibilidad y Medio Ambiente (2022) Escenarios Locales de Cambio Climático en Andalucía actualizados al 6° Informe IPCC

Dialesandro J, Brazil N, Wheeler S, Abunnasr Y (2021) Dimensions of thermal inequity: neighborhood social demographics and urban heat in the Southwestern US. Int J Environ Res Public Health 18(3):941

Diz-Mellado E, López-Cabeza VP, Roa-Fernández J, Rivera-Gómez C, Galán-Marín C (2023) Energy-saving and thermal comfort potential of vernacular urban block porosity shading. Sustain Cities Soc 89:104325

Dumlao A, Debbage N (2023) Examining the spatial proximity of vulnerable communities to cooling centers across the Texas Triangle Megaregion. Weather Climate Soc 15(4):1085–1098

Ebi KL, Capon A, Berry P, Broderick C, de Dear R, Havenith G, Malik A et al (2021) Hot weather and heat extremes: health risks. The Lancet 398(10301):698–708

Eisenman DP, Wilhalme H, Tseng C-H, Chester M, English P, Pincetl S, Dhaliwal SK et al (2016) Heat Death Associations with the built environment, social vulnerability and their interactions with rising temperature. Health Place 41:89–99

Frishman M, Uddin Khan H, Al-Asad M (2002) The Mosque: history, architectural development & regional diversity

Gabaldón-Leal C, Lorite IJ, Mínguez MI, Lizaso JI, Dosio A, Sanchez E, Ruiz-Ramos M (2015) Strategies for adapting maize to climate change and extreme temperatures in Andalusia, Spain. Climate Res 65:159–173

Ghafarian F, Wieland R, Nendel C (2022) Estimating the evaporative cooling effect of irrigation within and above soybean canopy. Water 14(3):319

González Santa Cruz F, Moral-Cuadra S, López-Guzmán T (2019) Analysis of the motivation and satisfaction of tourists in relation to an intangible cultural heritage: Fiesta de los Patios, Cordoba, Spain. Heritage Soc 12(1):26–40. https://doi.org/10.1080/2159032X.2020.1764306

Green H, Bailey J, Schwarz L, Vanos J, Ebi K, Benmarhnia T (2019) Impact of heat on mortality and morbidity in low and middle income countries: a review of the epidemiological evidence and considerations for future research. Environ Res 171:80–91

Guerrero JJ, Caceres F, Giménez F, Moreira JM (2016) Servicios de regulación climática aportados por la vegetación urbana a la Ciudad de Córdoba. 1ª Parte: Fundamentos y Metodología. REDIAM. Consejería de Medio Ambiente y Ordenación Del Territorio. Junta de Andalucía

Hayden MH, Wilhelmi OV, Banerjee D, Greasby T, Cavanaugh JL, Nepal V, Gower S et al (2017) Adaptive capacity to extreme heat: results from a household survey in Houston, Texas. Weather Climate Soc 9(4):787–799

Kim Y, Lee W, Kim H, Cho Y (2020) Social isolation and vulnerability to heatwave-related mortality in the urban elderly population: a time-series multi-community study in Korea. Environ Int 142:105868

Kjellstrom T, Butler AJ, Lucas RM, Bonita R (2010) Public health impact of global heating due to climate change: potential effects on chronic non-communicable diseases. Int J Public Health 55:97–103

König C (2024) Neighbourhood structure and environmental quality: a fine-grained analysis of spatial inequalities in urban Germany. Urban Stud 00420980231224224

Lee M-S, Park Y (2015) The courtyard as a microcosm of everyday life and social interaction. Architectural Res 17(2):65–74

Liu J, Varghese BM, Hansen A, Xiang J, Zhang Y, Dear K, Capon A et al (2021) Is there an association between hot weather and poor mental health outcomes? A systematic review and meta-analysis. Environ Int 153:106533

Loibl W, Vuckovic M, Etminan G, Ratheiser M, Tschannett S, Österreicher D (2021) Effects of densification on urban microclimate—a case study for the city of Vienna. Atmosphere 12(4):511

Lopez-Guzman T, Santa Cruz FG (2016) The Fiesta of the Patios: intangible cultural heritage and tourism in Cordoba, Spain

Martinelli L, Matzarakis A (2017) Influence of height/width proportions on the thermal comfort of courtyard typology for Italian climate zones. Sustain Cities Soc 29:97–106

Mexia T, Vieira J, Príncipe A, Anjos A, Silva P, Lopes N, Branquinho C et al (2018) Ecosystem services: urban parks under a magnifying glass. Environ Res 160:469–478

Nayak SG, Shrestha S, Sheridan SC, Hsu W-H, Muscatiello NA, Pantea CI, Hwang S-AA et al (2019) Accessibility of cooling centers to heat-vulnerable populations in New York State. J Transp Health 14:100563

Paz S, Negev M, Clermont A, Green MS (2016) Health aspects of climate change in cities with Mediterranean climate, and local adaptation plans. Int J Environ Res Public Health 13(4):438

Petruccioli A (2004) The courtyard house: typological variations over space and time. In: Courtyard housing. Taylor & Francis, pp 24–48

Phelan PE, Kaloush K, Miner M, Golden J, Phelan B, Silva H III, Taylor RA (2015) Urban heat island: mechanisms, implications, and possible remedies. Annu Rev Environ Resour 40(1):285–307

Pineo H, Rydin Y (2018) Cities, health and well-being

Priego-Gonzáles C, Baena L, Breuste J (2012) The courtyards of Cordoba in Spain. In: Bavarian Academy for Nature Conservation and Landscape Management (ed) Implementation of landscape ecological knowledge in European urban practice. Laufener Special Contributions, pp 67–72

Randazzo T, De Cian E, Mistry MN (2020) Air conditioning and electricity expenditure: the role of climate in temperate countries. Econ Model 90:273–287

Raviz SRH, Eteghad AN, Guardiola EU, Aira AA (2015) Iranian courtyard housing: the role of social and cultural patterns to reach the spatial formation in the light of an accentuated privacy. ACE: Arquitectura, Ciudad y Entorno

Redaelli G (2019) PAX—Patios de la Axerquía urban regeneration and social innovation in a heritage context. Built Heritage 3(1):91–104

Redaelli GA (2020) De la especulación a la cultura de la rehabilitación. La estrategia PAX-Patios de la Axerquía como medida antigentrificación en entornos patrimoniales. Kultur: Revista Interdisciplinària Sobre La Cultura de La Ciutat 7(13):91–116

Rohat G, Flacke J, Dosio A, Pedde S, Dao H, van Maarseveen M (2019) Influence of changes in socioeconomic and climatic conditions on future heat-related health challenges in Europe. Global Planet Change 172:45–59

Rojas-Fernández J, Galán-Marín C, Roa-Fernández J, Rivera-Gómez C (2017) Correlations between GIS-based urban building densification analysis and climate guidelines for Mediterranean courtyards. Sustainability 9(12):2255

Safarzadeh H, Bahadori MN (2005) Passive cooling effects of courtyards. Build Environ 40(1):89–104

Salisbury FB, Ross CW (1992) Plant physiology. Wadsworth Pub. Com., Inc., Belmont, California-USA

Sathyakumar V, Ramsankaran R, Bardhan R (2019) Linking remotely sensed Urban Green Space (UGS) distribution patterns and Socio-Economic Status (SES)—a multi-scale probabilistic analysis based in Mumbai, India. Giscience Remote Sens 56(5):645–669

Saura M, Beltran J, Pakseresht S (2017) A cross-cultural, comparative morphology-study of two composite cities: courtyard design in Barcelona and Kermanshah. In: Rethinking, reinterpreting and restructuring composite cities

Schüle SA, Gabriel KMA, Bolte G (2017) Relationship between neighbourhood socioeconomic position and neighbourhood public green space availability: an environmental inequality analysis in a large German city applying generalized linear models. Int J Hyg Environ Health 220(4):711–718

Singh N, Areal AT, Breitner S, Zhang S, Agewall S, Schikowski T, Schneider A (2024) Heat and cardiovascular mortality: an epidemiological perspective. Circ Res 134(9):1098–1112

Soflaei F, Shokouhian M, Shemirani SMM (2016) Traditional Iranian courtyards as microclimate modifiers by considering orientation, dimensions, and proportions. Front Architectural Res 5(2):225–238

Stone B Jr (2005) Urban heat and air pollution: an emerging role for planners in the climate change debate. J Am Plann Assoc 71(1):13–25

Suen I-S (2022) Assessment of Urban Heat Islands and Land Cover types in relation to vulnerable populations. Earth 3(2):733–747

Sun H, Jimenez-Bescos C, Mohammadi M, Zhong F, Calautit JK (2021) Numerical investigation of the influence of vegetation on the aero-thermal performance of buildings with courtyards in hot climates. Energies 14(17). https://doi.org/10.3390/en14175388

Taleghani M, Tenpierik M, van den Dobbelsteen A, Sailor D (2014) Heat in courtyards: a validated and calibrated parametric study of heat mitigation strategies for urban courtyards in the Netherlands. Sol Energy 103:108–124. https://doi.org/10.1016/j.solener.2014.01.033

Thompson R, Hornigold R, Page L, Waite T (2018) Associations between high ambient temperatures and heat waves with mental health outcomes: a systematic review. Public Health 161:171–191

Tung EL, Hawkley LC, Cagney KA, Peek ME (2019) Social isolation, loneliness, and violence exposure in urban adults. Health Aff 38(10):1670–1678

Turan İA, True EM (2023) The perception of public space of the elderly after social isolation and its effect on health. Ain Shams Eng J 14(3):101884

UN D (2015) World urbanization prospects: the 2014 revision. United Nations Department of Economics and Social Affairs, Population Division: New York, NY, USA, p 41

UNESCO (2003) Text of the convention for the safeguarding of the intangible cultural heritage

Wong T, Mackres E (n.d.) City-scale, city-relevant climate hazard indicators under 1.5° C, 2.0° C, and 3.0° C of global warming

Yuchi W, Sbihi H, Davies H, Tamburic L, Brauer M (2020) Road proximity, air pollution, noise, green space and neurologic disease incidence: a population-based cohort study. Environ Health 19:1–15

Zamani Z, Heidari S, Hanachi P (2018) Reviewing the thermal and microclimatic function of courtyards. Renew Sustain Energy Rev 93:580–595

Zhang N, Wang H, Gallagher J, Song Q, Tam VWY, Duan H (2020) A dynamic analysis of the global warming potential associated with air conditioning at a city scale: an empirical study in Shenzhen, China. Environ Impact Assessment Rev 81:106354

Addressing Energy Poverty in the Developed Countries: A Case Study from Japan

Zihao Zhu, Deanty Mulia Ramadhani, and Masachika Suzuki

1 Introduction

Energy and poverty are intertwined, as reflected in the underprivileged conditions in many developing and developed countries across the world. An estimated 2.7 billion people rely on traditional energy sources in their daily routines and lack access to clean cooking facilities (IEA 2024). Understanding the concept of energy poverty is a complex task, however, as it not only refers to energy access but also covers the linkages with economic, social, and environmental dimensions. The perspectives toward energy poverty are context-dependent as well among different countries and regions. How it is manifested could be different between developed and developing nations (Kez et al. 2024).

The United Nations (UN) outlined the alleviation of energy poverty within the Sustainable Development Goals (SDGs) through three fundamental pillars: energy and electricity access, electricity efficiency, and increased utilisation of renewable energy (Birol 2007). A wide range of existing studies regarding energy poverty literature are divided into two contexts. First, it is primarily focused on developing countries by identifying modern energy access and the impact of its accessibility towards social and economic sectors (Ahmad et al. 2014; Sadath and Acharya 2017; Banerjee et al. 2021), although quantifying energy poverty is challenging (Zhang et al. 2019). Meanwhile, in the context of developed countries, the discourse on energy poverty primarily focuses on energy and electricity affordability through an economic lens (Moore 2012; Okushima 2016). Despite less attention to the energy poverty issue in developed countries (Brunner et al. 2012), deliberations on energy

Z. Zhu (✉)
TMF Group, Tokyo, Japan
e-mail: zihao.zhu@tmf-group.com

D. M. Ramadhani · M. Suzuki
School of Global Environmental Studies, Sophia University, Tokyo, Japan

M. Okitasari et al. (eds.), *Equitable and Sustainable Cities*, Advances in 21st Century Human Settlements, https://doi.org/10.1007/978-981-92-2793-8_3

poverty are essential to explore effective policy formulations in developed countries as well (Gasser 2020; Kez et al. 2024).

This chapter first explores the concept of 'energy poverty' by explaining how the concepts and implications are discussed in different contexts, especially between the developing and developed countries. Then, it presents a case study in Japan where an analysis of energy poverty is conducted based on the data collected through a survey study to understand who is 'energy poor' and what factors are possibly leading to energy poverty in the country.

2 The Concept of Energy Poverty in Different Contexts

Addressing energy poverty is a pivotal variable in achieving a just transition towards an increase in fairness, efficiency, and inclusivity in the clean and renewable energy landscape (Li et al. 2024). In the sense of achieving a just transition, national and/or local governments must address energy poverty to ensure that communities can participate in and benefit from the shift to clean energy.

On the one hand, the prevalence of just transition is expected to alleviate energy poverty while improving economic and social well-being (Carfora and Scandurra 2024). On the other hand, the issue of willingness to pay and social acceptance of people in developing countries may hinder the energy shift from conventional to clean energy sources. Energy transition can bring about positive social impacts to the community only if it is implemented under effective measures and in a fair manner (Gonzalez-Eguino 2015; Spandagos et al. 2023). Considering the uncertain risks and profound socio-technical modifications, the transition to low-carbon energy may lead to social unacceptance (Saraji and Streimikiene 2023; Wsesterlund 2020).

In addition, the COVID-19 pandemic, which started at the beginning of 2020, caused enormous implications for household energy consumption and costs. Szamrej-Baran (2023) conducted a comparative study reviewing the risks of energy poverty among the European Union (EU) countries by examining how countries have dealt with energy poverty in terms of economic and energy expenses during the health crisis. The study measured the Index of Energy Poverty (IEP) to examine the significant impacts of the pandemic on the dynamics of energy poverty across the EU countries using three dimensions of variables: financial strain associated with residence or housing, social condition, household income, and housing conditions. However, the study faced several limitations in measures, such as the inability to provide the size of the phenomenon, as practical solutions to a number of people affected by energy poverty may be difficult to implement. According to Roberts et al. (2015), the estimated risks and the level of energy poverty in the EU can be assessed, yet as energy poverty is not unequivocally interpreted, it is challenging to construct long-term measures and draw a picture of energy poverty to other cases (Roberts et al. 2015).

Ongoing debates occur on the drivers of energy poverty both in developed and developing countries and how to determine the measurement of energy poverty at different countries' levels of development with different characteristics is considered a formidable challenge. On the other hand, there is a growing amount of literature on the measurement of the concept (Bouzarovski 2014; Kez et al. 2024). The contrasting implications of energy poverty in developed and developing countries have been widely discussed with various approaches. For example, the term 'fuel poverty' has evolved to address energy poverty in developed countries, especially recently industrialised countries, in recent studies. However, a conceptual gap exists within the evolving relationship between energy and fuel poverty (Primc et al. 2021). While energy poverty is perceived as the inability to provide modern energy services (Parajuli 2011), fuel poverty is regarded as the inability to pay for energy prices due to low income level (Castaño-Rosa et al. 2019).

In the context of developing countries, household electrification access becomes one of the energy poverty issues that leads to a consequent impact on the quality of life, particularly the health status of people in rural areas (Pereira et al. 2011). One of the illustrating methods to measure energy poverty at the household level in developing nations is called 'energy ladders', which note that low-income households have less energy consumption but have more household expenses compared to middle- and high-income households due to energy inefficiency and inequality (Barnes et al. 2004; Sovacool 2012). When households in developing countries lack access to modern energy services and have to rely on traditional energy sources, they are considered energy poor (Barkat et al. 2023). This situation is called a 'poverty trap' cycle that affects health and economic status (Drago and Gatto, 2023). In Ireland, cases of respiratory disease appear along with difficulty, poor quality, and energy-efficient housing (Charlier and Legendre 2023). Australia also faces a rising energy poverty level due to a deterioration in the health sector (Brown and Vera-Toscano 2021). Brazil, one of the countries that is perceived as the archetype of energy poverty, has made many efforts in this regard. One example is the attempt to reform energy by focusing on energy efficiency, rural electrification, and increasing per capita consumption of electricity (Pereira et al. 2010). The efforts were driven to encourage the decentralised energy in Brazil to create a sustainable, resilient society, and reduce the 'energy poor'.

As for the developed countries, European Union (EU) member states have encountered fuel poverty for years. A consensus was built on the fuel poverty measurement in 2000 in the EU based on the analysis of affordability to access basic needs among its citizens (Gordon et al. 2000). In 2009, the phenomenon of fuel poverty referred to as energy poverty was introduced by the EU Directive on common rules for the internal electricity market (2009/72/EC) (Official Journal of the EU 2009). The official definitions and drivers of fuel poverty vary across EU member states, while three countries in the EU region have formally defined what energy poverty is, illustrating the struggling households' access to adequate energy services (Thomson and Snell 2013). Energy community initiatives are prevalent and discussed as one of the government solutions to alleviate energy poverty through collaborative partnership and reduction of energy consumption and cost (Parreño-Rodriguez et al. 2023). Previous studies have discussed the extent to which collective action participation may tackle energy poverty, for example, through community training, forum discussions, community accountability sharing, and ownership (Button 2018; Fehr and Gächter 2000; Koukoufikis et al. 2023).

As for Japan, unlike in the EU, the issue of energy poverty is being considerably overlooked by its government, failing to establish official definitions and policy frameworks in place (Okushima 2016; Okushima 2019; Castano-Rosa and Okushima 2021). Community-based energy governance is not adequately discussed in the literature and is acknowledged as essential in addressing the issue of energy poverty in Japan (Koga et al. 2025). However, previous relevant studies about energy poverty in Japan exist. For example, Tabata et al. (2020) demonstrate that compared to other age groups, older people are more vulnerable to being exposed to fuel poverty as they stay at home longer, which leads to lower annual income and larger electricity use and cost (Tabata et al. 2020). Thus, exploring research on possible solutions to alleviate energy poverty in the energy transition has increasingly become an important research agenda in the efforts to cope with climate change in Japan.

3 Japan Case Study on Energy Affordability

As stated above, understanding the concept of energy poverty is complex, as it not only refers to energy access but also covers the linkages with economic, social, and environmental dimensions. This part of the article explores who is energy-poor and what leads to energy-poor based on the analysis of the data collected through a survey study conducted in Japan. The data was collected in February 2023 using a marketing survey company in Japan. The size of the samples was 3,000, targeting citizens from age 15 to 99 living in three different living environments: villages or small towns, the middle area refers to towns with less than 100,000 residents, and the large area refers to the urban area with more than 100,000 residents in total in Japan. In particular, the analysis examines different levels of energy affordability in these different living environments and possible factors contributing to the levels of energy affordability.

3.1 Income Differences Among Different Living Environments

Figure 1 illustrates the average ratings of people in different regions of residence on whether it is difficult to live on their current income. The higher the scores are, the harder their living is expressed by individuals.

People in the large area have the lowest scores, meaning that they experience the least difficult time living at their current income level. In contrast, those in the middle and small areas have higher scores, indicating that they experienced life as more difficult at their current income level. Despite there is no consensus on quantifying

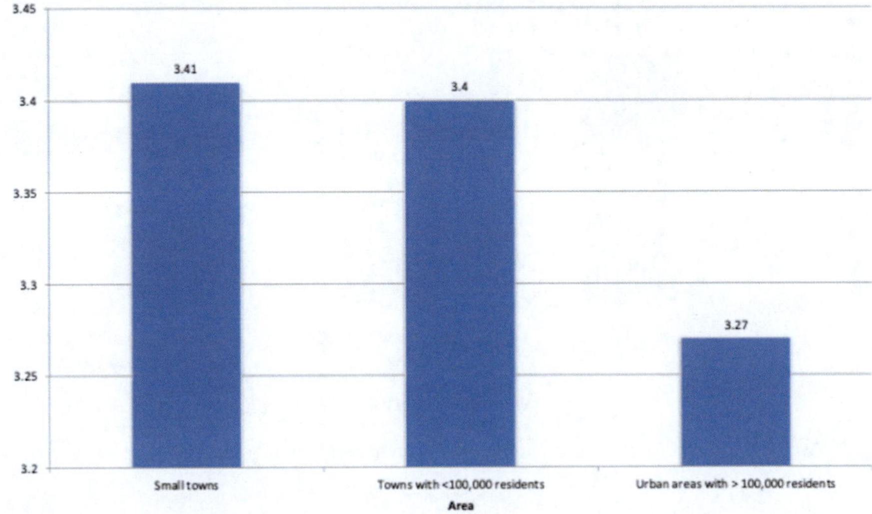

Fig. 1 Average income difficulty scores by area in Japan

energy poverty along with increasing debates on indicators and variables to identify energy poverty, these results extend the finding of Kez et al. (2024) by demonstrating that the household income level has an influence on the vulnerability to the energy poverty level, regardless of developed or less developed countries. Particularly, those with higher income difficulty scores tend to struggle to allocate sufficient funds for energy expenses which leads to a higher risk of energy poverty.

3.2 Differences in Perceived Cost of Energy Bills Among Different Living Environments

The next question is whether the scale of the living environment makes a difference in how people feel about the overhead of their energy bills. The images in Fig. 2 have similarly inverted scales to those in Fig. 1.

Figure 2 shows the average ratings (inverted scale) of people in the different residential areas on whether their electricity, gas, and other utility bills are perceived as excessively high. The numbers on each bar indicate the average rating for the corresponding area. We can see that the scores are notably high within all three areas in this figure. People living in the smaller areas have the highest scores, which means that the people living in the area are more sensitive to energy expenses compared to the people living in the large or middle areas. This could be because incomes are considerably higher in cities, posing less vulnerability (Lowans et al. 2023). These findings lend further support to the studies conducted by Okushima (2016, 2019) using the multi-dimensional energy poverty index (MEPI) in Japan to identify

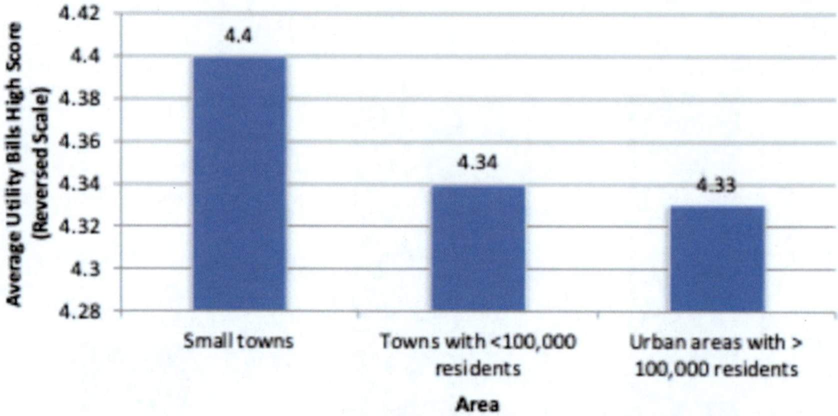

Fig. 2 Average ratings of utility bills among the different residential areas

energy-poor households using indicators such as energy costs, income, and energy efficiency. The index can be considered as an alternative method for evaluating energy poverty in Japan, which identifies the energy consumption and types of energy used at the household level, not limited to regional-based and socio-demographic analysis, but also identifies energy vulnerability factors, like climatic factors (Kahouli and Okushima 2021). The results of this study suggest that even after accounting for these indicators, residents in villages and small towns appear to perceive their utility bills as more difficult than those in larger areas. This reflects that the above indicators may not capture how people perceive the energy burden but also indicate other factors that may contribute to the distinction in perceived energy burdens. It shows that different regions, particularly with climatic differences, result in diverse energy consumption levels. A positive relationship was established between household income level and the spatial condition of the region, including the population age group and the scale of area (Okushima and Simcock 2024).

3.3 Differences in Utility Expenses Among Different Living Environments

Next, we examine the differences in energy or utility expenses among the three different areas. Figure 3 depicts how much people in the three areas spend annually on electricity, gas, kerosene, and gasoline. It is noted that the energy price could differ greatly among different regions in Japan, as the energy market is liberalised in the country, and the energy price could depend on differences in cost among energy producers and retailers.

As seen in Fig. 3, electricity is the largest part of the utility bill, regardless of the region. People in small areas spend more on utility bills than people in medium and

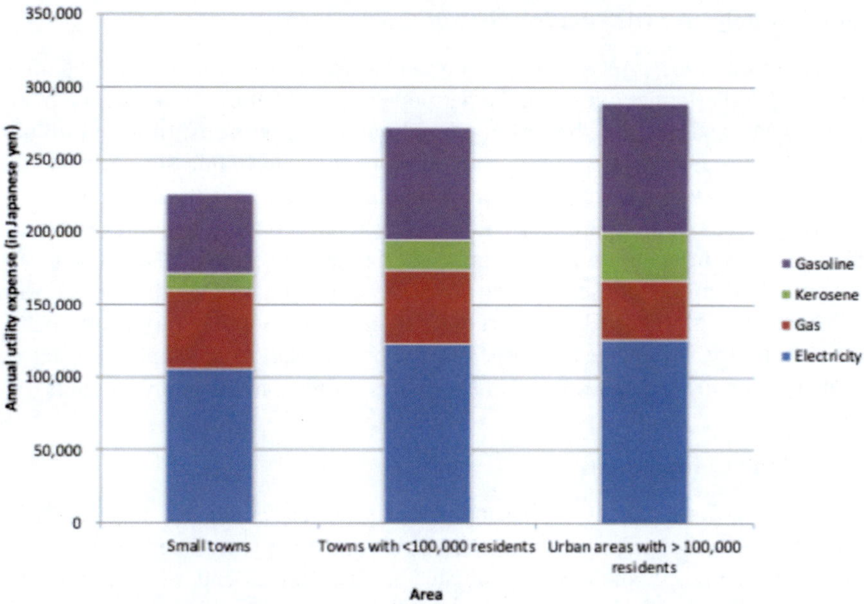

Fig. 3 Average annual utility expenses by area in Japan

large areas, with gasoline and kerosene use higher than other utility sources. Despite the adverse impacts of kerosene, people in villages or small towns predominantly use it as a lighting source as it costs less than other energy sources to meet the needs of rural communities for cooking and other purposes (Rehman et al. 2005; Reddy et al. 2000; Kojima 2021).

Based on the trends observed in the above figures, we maintain that the residents in the small towns have higher levels of energy poverty than those in the medium-sized towns and urban areas. This can be attributed to the larger reliance on gasoline and kerosene in small areas, which is considered less efficient and has higher costs in the long term. Infrastructure limitations in small areas also impact residents' access to cost-effective energy sources compared to large and middle areas. Therefore, it is necessary to understand that energy poverty cannot be solely determined by expenditure but also by looking at energy affordability. Despite having the lowest utility bills, residents in large areas may perceive energy costs to be burdensome. This result echoes the established energy expenditure-based approach that perceives the Low Incomes and High Costs (LIHC) indicator where the households in the low-income group are considerably energy-poor (Hills 2012; Igawa 2022).

There are various reasons why people in the small area spend more on their energy bills. While the reasons could vary depending on different circumstances and contexts, the following could be considered as key reasons for the difference in the energy expenses between the large and small areas:

Mature Supply and Distribution Networks

In large cities, the supply and distribution networks of utilities may be more centralised and efficient due to the higher population densities. Transmission losses for electricity and gas may be lower, which may reduce the cost of these utilities. Large populations and the resulting economies of scale make infrastructure in cities often cheaper than in rural areas. Also, less populated areas mean that residents must travel long distances to get around. The accessibility to economic activities and financial inclusion in infrastructure development and energy investment also influence the rural area to be more exposed to energy poverty compared to the urban area (Wang et al. 2024). In addition, in the context of traditional infrastructure providers, such as water, electricity, gas and network operators or transportation companies, this means that their costs must be allocated to fewer users. In turn, individual users must pay more.

Result of the Competitions

In large cities, there may be more suppliers of electricity and gas, and competition among them may lower prices. Electricity grids and energy providers, for instance, tend to start expansion in densely populated areas before extending the coverage to less densely populated ones like rural areas (Khandker et al. 2010).

Energy Efficiency

The characteristics of the small and rural area communities that serve fewer populations lead to challenges in energy efficiency delivery (Shoemaker et al. 2018). The small and rural areas are distinct in energy infrastructures and technologies from modern facilities in large cities with a larger scale of energy efficiency. The term energy efficiency has also been aligned with the concept of urban planning as a pivotal component through sustainable and green architecture and modern efficient public transportation (Umoh et al. 2024).

3.4 Socio-Economic Factors Associated with Energy Poverty

Figure 1 indicates that people in the middle and small areas feel that life is more difficult than in the large areas based on their current income level. Here, we demonstrate a model with the dependent variable: 'Do you find it difficult to live on your current income'? The independent variables in this model are annual income, family structure, number of rooms in the house, and education. The following are some explanations about each variable:

Annual Income Including Taxes. Annual income, including household taxes, ranges from less than 1 million yen to more than 20 million yen.

Family Structure. Family structure ranges from a single-person household to three generations.

Number of Rooms. Number of rooms ranging from one room to three or more rooms.

Education Level. The education level ranges from less than a high school diploma to more than a master's degree.

Since logistic regression requires a dichotomous variable, logistic regression was performed by coding 1 (Not hard at all), 2 (Not very hard), 3 (Can't say either way) as 0 and 4 (Hard) 5 (Very hard) coded as 1, logistic regression was performed. The results were as follows:

	Coeff	Std err	z	P > |z|
Const	−0.0851	0.043	−1.977	0.048
X1 (Education level)	−0.1255	0.044	−2.872	0.004
X2 (Income including household taxes)	−0.6151	0.053	−11.619	0.000
X3 (Family structure)	0.3685	0.049	7.538	0.000
X4 (Number of rooms)	0.0065	0.047	0.138	0.890

The results show that income and family structure are associated with the level of difficulty of living. The effect of education is also statistically significant, albeit to a slightly lesser extent. In fact, several existing studies have provided several examples of the impacts of energy poverty on access to education (Toman and Jemelkova 2003; Cabraal et al. 2005). For example, some research indicates that a lack of access to education has a role in deepening energy poverty (Acharya and Sadath 2019). However, according to Banerjee et al. (2021), the energy poverty-education nexus has not yet been fully empirically tested.

4 Concluding Remarks

This chapter demonstrated that small or rural areas are more prone to energy poverty than urban areas. Less affordable energy services in the small area communities were illustrated by examining the associations between the living environment and the people's feelings toward living and energy expenses. In addition, the lower income yet higher energy expenditure in the small area demonstrated how energy poverty could play an impactful role in the people's lives in the small and local communities. Further study on the perspectives of energy poverty assessment is necessary to enrich ongoing research on this topic and to deepen our understanding of a better energy poverty evaluation approach and model in different socio-economic contexts and settings.

5 Declaration of Competing Interests

The research described in this article is supported by the MEXT/JSPS KAKENHI Grant Number JP20K01917 as well as a joint research project between UNU-IAS and Sophia University under the Grant for Global Sustainability (GGS).

References

Acharya RH, Sadath AC (2019) Energy poverty and economic development: household-level evidence from India. Energy Build 183:785–791. https://doi.org/10.1016/j.enbuild.2018.11.047

Ahmad S, Mathai MV, Parayil G (2014) Household electricity access, availability, and human well-being: evidence from India. Energy Policy 69:308–315. https://doi.org/10.1016/j.enpol.2014.02.004

Banerjee R, Mishra V, Asfaw A (2021) Energy poverty, health, and education outcomes: evidence from the developing world. Energy Economics 101:105447. https://doi.org/10.1016/j.eneco.2021.105447

Barkat K, Alsamara M, Mimouni K (2023) Can remittances alleviate energy poverty in developing countries? New evidence from panel data. Energy Economics 119:106527. https://doi.org/10.1016/j.eneco.2023.106527

Barnes D, Krutilla K, Hyde W (2004) The urban household energy transition: energy, poverty, and the environment in the developing world. World Bank Policy Research Working Paper No. 6903. https://www.esmap.org/sites/esmap.org/files/Rpt_UrbanEnergyTransition.pdf

Birol F (2007) Energy economics: a place for energy poverty in the agenda? Energy J 28:1–6

Bouzarovski S (2014) Energy poverty in the European Union: landscapes of vulnerability. Wiley Interdisc Rev Energy Environ 3(3):276–289. https://doi.org/10.1002/wene.89

Brown H, Vera-Toscano E (2021) Energy poverty and its relationship with health: empirical evidence on the dynamics of energy poverty and poor health in Australia. SN Bus Econ 1(10):139. https://doi.org/10.1007/s43546-021-00149-3

Brunner K, Spitzer M, Christanell A (2012) Experiencing fuel poverty. coping strategies of low-income households in Vienna/Austria. Energy Policy 49:53–59. https://doi.org/10.1016/j.enpol.2011.11.076

Cabraal A, Barnes DF, Agarwal SG (2005) Productive uses of energy for rural development. Annual Rev Environ Resources 30:117–144. https://doi.org/10.1146/annurev.energy.30.050504.144228

Carfora S, Scandurra G (2024) Boosting green energy transition to tackle energy poverty in Europe. Energy Res Soci Sci 110:103451. https://doi.org/10.1016/j.erss.2024.103451

Castaño-rosa R, Solís-guzmán J, Rubio-bellido C, Marrero M (2019) Energy & buildings towards a multiple-indicator approach to energy poverty in the European Union: a review 193:36–48. https://doi.org/10.1016/j.enbuild.2019.03.039

Cong S, Nock D, Qiu Y, Xing B (2022) Unveiling hidden energy poverty using the energy equity gap. Nat Commun 13:2456. https://doi.org/10.1038/s41467-022-30146-5

Drago C, Gatto A (2023) Gauging energy poverty in developing countries with a composite metric of electricity access. Utilities Policy 81:101486. https://doi.org/10.1016/j.jup.2022.101486

Fidelis N, Giannini M, Aure M (2010) Rural electrification and energy poverty: empirical evidences from Brazil 14:1229–1240. https://doi.org/10.1016/j.rser.2009.12.013

Gasser P (2020) Review article. Energy Policy 139:111339. https://doi.org/10.1016/j.enpol.2020.111339

González-eguino M (2015) Energy poverty: an overview 47:377–385. https://doi.org/10.1016/j.rser.2015.03.013

Gordon D, Adelman L, Ashworth K et al (8 more authors) (2000) Poverty and social exclusion in Britain. Research Report. Joseph Rowntree Foundation, York. https://eprints.whiterose.ac.uk/73358/

Hills J (2012) Getting the measure of fuel poverty. Final report of the fuel poverty review. Centre for Analysis of Social Exclusion, London School of Economics and Political Science, London

IEA (2024) SDG7: data and projections, IEA, Paris https://www.iea.org/reports/sdg7-data-and-projections. Licence

Igawa M (2022) Empirical analyses of energy poverty at the household level empirical analyses of energy poverty at the household level. Kyushu University Institutional Repository. https://hdl.handle.net/2324/5068205

Kahouli S, Okushima S (2021) Regional energy poverty reevaluated:aA direct measurement approach applied to France and Japan. https://www.sciencedirect.com/science/article/pii/S0140988321003777

Koga H, Bouzarovski S, Petrova S (2025) Unsettling mainstream academic debates on community-based energy governance: exploring the Japanese experience. Renew Sustain Energy Rev 207:114994. https://doi.org/10.1016/j.rser.2024.114994

Kez AD, Foley A, Lowans C, Furszyfer D, Rio D (2024) Energy poverty assessment: Indicators and implications for developing and developed countries. Energy Convers Manage 307:118324. https://doi.org/10.1016/j.enconman.2024.118324

Khandker SR, Barnes DF, Samad HA (2010) Energy poverty in rural and urban India. Are the energy poor also income poor? Policy Research Working Pape 5463:1–38. http://documents.worldbank.org/curated/en/920901468259774557/pdf/WPS5463.pdf

Kojima M (2021) Primary Household energy for cooking and heating in 52 developing economies. Primary Household Energy for Cooking and Heating in 52 Developing Economies, June. https://doi.org/10.1596/35947

Li G, Jiang M, Yuan Y, Chen X, Fu D (2024) Heliyon The nexus of poverty energy in China' s industrial productive efficiency and energy transition in digital economy. Heliyon 10(14):e34247. https://doi.org/10.1016/j.heliyon.2024.e34247

Lowans C, Foley A, Furszyfer D, Rio D, Sovacool K (2023) Towards more equitable energy transitions in low-income households: an integrated analysis of energy and transport poverty in Northern Ireland. Energy Convers Manage 291:117337. https://doi.org/10.1016/j.enconman.2023.117337

Moore R (2012) Definitions of fuel poverty: implications for policy. Energy Policy 49:19–26. https://doi.org/10.1016/j.enpol.2012.01.057

Official Journal of the European Union (2009) Directive 2009/73/EC of the European Parliament and the Council, July 2009

Okushima S (2016) Measuring energy poverty in Japan, 2004–2013. Energy Policy 98:557–564. https://doi.org/10.1016/j.enpol.2016.09.005

Okushima S (2019) Energy & buildings understanding regional energy poverty in Japan: a direct measurement approach. Energy Build 193:174–184. https://doi.org/10.1016/j.enbuild.2019.03.043

Okushima S, Simcock N (2024) Double energy vulnerability in Japan. Energy Policy 191:114184. https://doi.org/10.1016/j.enpol.2024.114184

Parajuli R (2011) Access to energy in Mid/Far west region-Nepal from the perspective of energy poverty. Renew Energy 36(9):2299–2304. https://doi.org/10.1016/j.renene.2011.01.014

Parreño-Rodriguez A, Ramallo-González AP, Chinchilla-Sánchez M, Molina-García (2023) Energy & buildings community energy solutions for addressing energy poverty: a local case study in Spain 296. https://doi.org/10.1016/j.enbuild.2023.113418

Pereira MG, Freitas MAV, da Silva NF (2010) Rural electrification and energy poverty: empirical evidences from Brazil. Renew Sustain Energy Rev 14(4):1229–1240. https://doi.org/10.1016/j.rser.2009.12.013

Ramallo-gonz AP, Parre A, Molina-garcía A (2023) Energy & buildings community energy solutions for addressing energy poverty: a local case study in Spain 296. https://doi.org/10.1016/j.enbuild.2023.113418

Reddy AKN, Smith KR, Williams RH (2000) Rural energy in developing countries. World Energy Assessment: Energy and the challenge of sustainability, Chapter 10. United Nations Development Programme

Rehman IH, Malhotra P, Pal RC, Singh PB (2005) Availability of kerosene to rural households: a case study from India. Energy Policy 33(17):2165–2174. https://doi.org/10.1016/j.enpol.2004.04.014

Roberts D, Vera-toscano E, Phimister E (2015) Fuel poverty in the UK: is there a difference between rural and urban areas? Energy Policy 87:216–223. https://doi.org/10.1016/j.enpol.2015.08.034

Sadath AC, Acharya RH (2021) Access to modern energy and human development in India: has government policies paid off? Int J Energy Econ Policy 11(3):S432–442

Saraji MK, Streimikiene D (2023) Challenges to the low carbon energy transition: a systematic literature review and research agenda. Energ Strat Rev 49:101163. https://doi.org/10.1016/j.esr.2023.101163

Shoemaker M, Gilleo A, Ferguson J (2018) Reaching rural communities with energy efficiency programs, 54

Sovacool BK (2012) Energy for sustainable development the political economy of energy poverty: a review of key challenges. Energy Sustain Dev 16(3):272–282. https://doi.org/10.1016/j.esd.2012.05.006

Spandagos C, Angel M, Rea T (2023) Energy poverty prediction and effective targeting for just transitions with machine learning 128. https://doi.org/10.1016/j.eneco.2023.107131

Szamrej-baran I (2023) ScienceDirect ScienceDirect Assessing the effects of the Covid-19 pandemic on energy poverty in EU countries the Index of Energy Assessing the effects of using the Covid-19 pandemic on Poverty energy poverty in EU countries Izabela. Proc Comput Sci 225:4187–4195. https://doi.org/10.1016/j.procs.2023.10.415

Thomson H, Snell C (2013) Quantifying the prevalence of fuel poverty across the European Union. Energy Policy 52:563–572. https://doi.org/10.1016/j.enpol.2012.10.009

Toman MA, Jemelkova B (2003) Energy and economic development: an assessment of the state of knowledge. Energy J 24(4):93–112. https://doi.org/10.5547/ISSN0195-6574-EJ-Vol24-No4-5

Umoh AA, Ohenhen PE, Chidolue O, Ngozichukwu B, Fafure AV, Ibekwe KI (2024) Incorporating energy efficiency in urban planning: a review of policies and best practices. Eng Sci Technol J 5(1):83–98. https://doi.org/10.51594/estj.v5i1.729

Zhang D, Li J, Han P (2019) A multidimensional measure of energy poverty in China and its impacts on health: an empirical study based on the China family panel studies. Energy Policy 131:72–81. https://doi.org/10.1016/j.enpol.2019.04.037

Housing Cooperative: A Force to Be Harnessed for Climate Change Mitigation and Adaptation

Ajibola Anthony Akanji and Paul Hudson

1 Introduction

1.1 Housing and Climate Change

It is projected that by 2050 it is projected that ~68% of the global population will live in urban areas (United Nations Department of Economic and Social Affairs 2018) combined with a projected global population of 9.5 billion by 2050 (compared to the current global population of 8 billion (United Nations 2022). This growth in population will, therefore, be increasingly urban and need suitable accommodation that is fit for purpose today and into the future.

IPCC (2007) found that Global greenhouse gas (GHG) emissions from buildings accounted for 21% of global GHG emissions out of which 24% were direct emissions (i.e. on-site emissions), 57% were indirect emissions (i.e. off-site emissions mainly related to heat and electricity production) and 18 per cent were embodied emissions (i.e. those related to the construction of the building). This is a major contribution towards global climate change and needs to be addressed, especially as new housing estates will be constructed to fit this growing need. In expanding the housing stock, we not only need to be concerned with the GHG emissions from the construction and continued existence of the grown housing stock but also how the expanded housing stock impacts how climate change is experienced. For example, expanding urban areas potentially increases the sealed surface area which can lead to more flooding as runoff cannot permeate into the ground as easily (Pistocchi 2017), while the loss of (urban) green spaces can increase the problems of urban heat islands magnifying the problem of extreme heat (Gago et al. 2013). Therefore, as our global housing stock is

A. A. Akanji (✉)
Faculty of Law, Lead City University, Ibadan, Oyo State, Nigeria

P. Hudson
Department of Environment and Geography, University of York, York, UK

M. Okitasari et al. (eds.), *Equitable and Sustainable Cities*, Advances in 21st Century Human Settlements, https://doi.org/10.1007/978-981-92-2793-8_4

expanded to meet increases in population this expansion must be managed to prevent it from becoming maladaptive, e.g. where we increase GHG emissions consistently (worsening the magnitude of climate change) or magnify the consequences of climate change (Sun and Weng 2024). For instance, as urbanisation happens in an unplanned and unregulated manner natural capital assets, such as forests, that help to act against climate change, are lost to unsustainable housing schemes (Onanuga et al. 2022).

While expanding the global housing stock in a climate-smart way is important, it must be noted that the rate of urbanisation and population growth will manifest differently across the globe. For instance, over half of the global population increase by 2050 is expected to occur in Africa (United Nations 2022); where the population of sub-Saharan Africa is expected to double by 2050 (United Nations 2022). Similarly, urbanisation around Africa is at an all-time high record (El-Bouayady and Radoine 2023) as is inward migration from rural to urban areas. A good example of this pressure is Nigeria which while currently the world's seventh largest country by population is projected to become the third-largest country by population in the world shortly before 2050. Moreover, this pressure of this sustained population growth places on Nigerian housing is already being experienced, especially in the major cities of Lagos, Ibadan and Abeokuta (Abiodun and Salami 2023). This pressure can result in unsustainable housing schemes, and expansion of relevant infrastructures, and the loss of valuable natural habitat as land use land cover (LULC) is changed at potentially such scale whose full environmental consequences may not be clear and quantifiable. For example, consider the loss of massive habitats along the *Lekki* corridor of Lagos, the former *Igbo efon* and *Igbo elerin,* respectively, forest of buffalos and forest of elephants. In the aforementioned areas, the large expanse of forest land lost to housing and supporting infrastructure contributes to environmental degradation.

Addressing how we can expand the housing stock without undermining our overall environmental and social goals is important and part of the Nigerian government's policy agenda. For instance, Nigeria has committed to reducing emissions by 20% unconditionally under a business-as-usual scenario, with additional reductions contingent on further international support, by 2030 (Noah 2024). Moreover, there is a suite of further Nigerian policy documents supporting joint action on climate change mitigation and adaptation. For example, the 2011 National Adaptation Strategy and Plan of Action on Climate Change for Nigeria (NASPA-CCN), established in 2011, was the country's first comprehensive adaptation policy framework; the 2012 National Climate Change Policy Response and Strategy (NCCPRS); the 2021 National Climate Change Policy (NCCP) for 2021–2030, which among other topics has a focus on sustainable land-use practices, climate-proof infrastructure and capacity building (Noah 2024). The policy articulates strategic objectives, including enhancing adaptation and mitigation measures.

Therefore, under this policy scope there is scope for innovative mechanisms from both within and without the Nigerian government to address the impasse of deficit in housing stock, current unsustainable housing models and the future sustainability of the Nigerian people. An intriguing policy option to explore is the use of housing cooperatives as a policy instrument for altering and constraining the incentives at the level of individual properties and their owners. This chapter will explore how the concept of housing cooperatives can be used in Nigeria to support the twin goals of climate change mitigation and adaptation, and how novel and/or emerging technologies may be able to support the ability of housing cooperatives to be transformative within the Nigerian context.

1.2 Housing Cooperatives

Cooperatives are autonomous associations of persons voluntarily united to meet their common economic, environmental, social and cultural needs through jointly owned and democratically controlled enterprises (International Cooperative Alliance 2025). The International Cooperative Alliance states that cooperatives are guided by principles and values: voluntary and Open Membership; Democratic Member Control; Member Economic Participation; Autonomy and Independence; Education, Training and Information; Cooperation among Cooperatives; and Concern for Community. This could also be expressed through the values of: Self-help; Self-responsibility; Democracy; Equity; and Solidarity. While not expressed in directly the same language and theoretical concepts there is an overlap of these principles and values with those that are considered to underpin successful community-based efforts to manage collective resources, see e.g. Ostrom (1990), as well as community-based efforts for climate change mitigation and adaptation, e.g. Schipper et al. (2014).

Within this vein, housing cooperatives are a not-for-profit mechanism that pools micro resources for housing projects. Individuals do not own houses, rather they are members of collectives that own houses, hence becoming joint owners or shareholders of the collective property. Therefore, there may be a tendency towards relatively dense housing units (e.g. apparently blocks, condominiums). Housing cooperatives are entities built on social values that ideally embed the philosophy of long-term viability, protection of member's assets and fulfilment of social goals which must be attained to secure the housing needs of their members.

Importantly, current global experiences with housing cooperatives show that they have the tendency to prioritise environmental factors and sustainability (Marckmann et al. 2012). For example, the non-profit and social value orientation of housing cooperatives can allow them to explore innovative building designs, construction materials, innovative maintenance, methods of building renovation and innovative management of community spaces (Balmer and Gerber 2018) that might otherwise not be undertaken due to a perception that they are not commercially viable. However, care must be taken within this statement as it is unclear to what extent it is the nature of how cooperatives contain incentives and perspectives that result in this outcome vs. the selective participation of individuals into such schemes that drive this outcome. Moreover, the cooperative's goal of providing social or collective value rather than private commercial value can also render the housing provided more affordable or at a higher level of quality while the collective ethos can be maintained, and the actions of the residents remain aligned. This is due to the different objectives that the cooperative management has.

However, while housing cooperatives appear well established in many countries of the Global North, the reserve appears to be the case in the many countries of the Global South, particularly in Africa where the only visible models are concentrated in Kenya, South Africa, Tanzania, Uganda, and Zimbabwe (Ganapati 2014) and lately Egypt. The practice of housing cooperatives is still in its infancy in Nigeria, this inability to effectively replicate cooperative housing in Nigeria may be in part because governing laws on cooperatives as currently constituted often result in unclear property rights and other legal rights that lead to frail cooperatives, and sub-optimal utilisation of their resources (Akanji 2022). While the concept is in its infancy in Nigeria (limiting the current scope to which it can be used as an instrument to address multiple social goals, e.g. affordable housing, lower embodied GHG emissions, inbuilt passive flood-proofing), the infancy provides the opportunity to influence a model through which housing cooperatives will develop and more strongly embed wider environmental change concerns.

Housing cooperatives can, arguably, be found in two forms. The first is a cooperative focused on land acquisition and sharing. Cooperatives that focus on these activities focus on the acquisition of land, which is then shared out to its members on agreed terms, for example, according to each member's shareholding, or subscription to a particular acquisition, or based on agreed terms in accordance with the provisions of the cooperative's bylaws or charter. Individual members of the cooperatives who have acquired plot(s) from the cooperatives proceed to build their individual houses on their acquired plots.

On some occasions, such houses are built according to specified patterns as earlier agreed by members of the cooperative society; under such circumstance, all houses built on the land acquired by the cooperatives adopt the same pattern or structure. On other occasions, the discretion of individual members dictates the pattern or structure of each house. Either way, general facilities such as roads, water, sewage systems, etc. are shared. This approach supports collectivism and standardisation.

Often, to undertake the endeavour of building their individual houses on their respective plots, some of these members do approach their cooperative society for house-building loans. Such loans are only available to financial members of the cooperative society—aside they serve as one of the control mechanisms of a cooperative society on its members.

The second is real property cooperatives, which rather than purchasing land to develop these cooperatives purchase existing buildings (either to renovate or use in situ). The resulting property is then jointly owned by all members of the cooperative society.

In both cases it must be noted that 'ownership' is different from 'possession', thus the cooperative society owns the building, however it may not necessary be in possession. Invariably, a member occupying such property does so as a tenant, and not as an owner. This can create potential conflicts between the two if not correctly managed and clear records of ownership, possession and the resulting obligations.

While housing cooperatives may, currently, be relatively rare in Nigeria, other forms of cooperative can be found in the form of agricultural, staff or financial cooperatives. The pooling of joint ownership, capacity and responsibility is the key driver through which cooperatives offer a range of opportunities to enhance environmental action. This potential in turn is driven by the observation that cooperatives have at their core a goal of providing social value rather than a consistent focus on profit maximisation (Sacchetti and Tortia 2014), in changing their core objective away from a traditional commercial perspective there are a range of options offered by cooperatives to act as an economic institution. An economic institution is a structure that acts to constrain and direct the behavioural incentives into a direction that is deemed to be 'acceptable' and in accordance with the institution's core values (Schmid 2008). Therefore, if a housing cooperative's core values and social objectives are aligned with environmental sustainability, we can create new opportunities for both climate change mitigation and climate change adaptation.

2 Housing Cooperatives: Climate Change Mitigation Opportunities

2.1 Building and Behavioural Regulation

The previous section noted that housing cooperatives can be seen as an economic institution because the nature and objectives instilled by the cooperative charter or agreements focus behaviour down certain socially orientated pathways. Therefore, this creates scope for linking the expectations of the cooperative to social and environmental outcomes and behavioural change that aid in achieving both sustainable development and climate change mitigation. While overall structural economic change and transition to a new economic model is likely required to truly mitigate climate change, behavioural change can contribute towards overall climate change mitigation (Newell et al. 2022).

For instance, it can be possible that when agreeing to take part in a housing cooperative, the participant agrees to abide by the cooperative's 'sustainability plan'. For example, housing cooperatives in Canada can have a template environmental sustainability policy based on that issued by a national trade body (Cooperative Housing Federation of Canada 2025). The adoption of such a code of practice can help promote leadership in two ways. The first is that the charter can incentivise the creation of various environmental initiatives, e.g. by introduction 'building expectations' for properties that can be included in the property portfolio or be encouraged to retrofit existing properties. This can be self-made regulations on energy and water efficiency standards to reduce consumption of natural resources, there can be expectations on nature coverage (e.g. each property must have access to a certain amount of green or blue space) for mental wellbeing with slow drip irrigation combined with native plants for the local habitat to improve local biodiversity. This can also be extended to the building materials used during the retrofit process as well (e.g. a focus on materials with a low level of embodied GHG emissions). This could be manifested through installing processes like government procurement processes where points are awarded based not only on the commercial rationale of expenditure but also the social and environmental value provided, see e.g. Crown Commercial Service (2025). While these self-imposed regulations can constrain the ability of the cooperative's decision-makers concerning the built environment, a similar code of conduct can also be established for the members of the cooperative as well.

The foregoing examples may be adapted to Nigeria although with a great deal of caution as they must be adapted to local peculiarities and contexts, which in some instances may not be uniform. For instance, there are different environmental conditions across Nigeria which dictate environmental designs, in addition to the diverse cultural and religious beliefs of the different peoples that are dominant in the different locations across the country. Another important source of diversity that may influence design and behavioural patterns in Nigeria is the legal system. While the 1999 Constitution of the Federal Republic of Nigeria and the Land Use Act are of general application throughout the country, the legislature in each state (state houses of assembly) reserves the discretion to make subordinate or addendum regulations to the relevant federal legislation. Many of such subordinate or addendum legislation are specific, hence differ in meaning and interpretation. Therefore, there will have to be a degree of differentiation across the federal states rather than a singular uniform model.

The ability of self-regulation is not to be underestimated, as it can act as can cause people to internalise the social consequences of their actions. This is the very essence of cooperative bylaws, a set of internal guiding rules made by the members of each cooperative society. This can be achieved because traditional economic policy and legal regulations assume that people are rational and logically respond to incentives and seek to maximise their self-interest unless forced to consider the altruistic interests of others. However, other perspectives such as that proposed by Zelizer (2012) imply that people seek 'meaningful' economic transactions and social interactions meaning that instead action is relational rather than transactional. The socially orientated nature of a cooperative allows society to maximise the potential offered by this more relationship-influenced concept of behaviour due to the people who are willing to undertake and support these meaningful interactions to self-select into membership of such a focused cooperative.

2.2 Energy Efficiency and Production

Cabeza et al. (2022) note that the majority of GHG emissions related to housing come from energy consumption related to heating/cooling and electricity production. Moreover, it must also be noted that reliable access to sustainable energy produces a crossover with the Sustainable Development Goals. Therefore, there is a potential overlap of two sources of social value that housing cooperatives could be used to develop and direct Nigeria's development pathway.

One of the ways through which housing cooperatives could act in this joint direction is through the creation, merger, or collaboration with 'energy cooperatives' for the properties within their portfolio (like how many Nigeria cooperatives also are de facto financial cooperatives as well, as mentioned in (Chilokwu et al. 2019). and can help scale energy activities as they mature into active generation, transmission and distribution (Hufen and Koppenjan 2015). Energy cooperatives have proven to be successful globally, e.g. see Germany (Punt et al. 2022) or Bolivia (Lillo et al. 2021), especially when tied to the production of renewable energy.

Succinct to state that while the use of alternative and sustainable sources energy is on the rise in Nigeria, the idea of energy cooperative like their housing cooperative counterpart is in its infancy. Notwithstanding, there is visible growth in the number of persons and institutions subscribing to the idea of energy cooperatives in the country. These Nigerian energy cooperatives have emerged primarily due to a mixture of high prices and energy insecurity (Monyei et al. 2022). The pressures of energy insecurity can be seen in Boateng et al. (2020) who note that 73% of their study participants reported physical energy insecurity (e.g. resulted to open fire/candle/stoves for lighting or the use of unsafe energy sources that increased exposures to potentially toxic gases). Moreover Boateng et al. (2020) noted that 88% had no access to any source of energy for cooking or lighting at least once in the last four weeks. The respondents in Boateng et al. (2020) then reported several reasons for why this occurred which can be mostly summarised as either not having access to replacement energy sources or the resources to pay for access to energy. Pokubo and Al-Habaibeh (2023) note that another source of energy insecurity is that the grid infrastructure is poorly managed rendering a higher reliance on conventional energy resources. One of the most profound contributions of energy insecurity to housing deficit and unsustainable housing in Nigeria is the restricted propensity of investors to embark on the construction of high-rise buildings due to the concern about how to source energy to power essential facilities expected to be within such buildings.

Pokoby and Al-Habaibeh (2023) note that as a way of addressing these challenges, the Nigerian government introduced a 'Renewable Energy Master Plan' to help transition towards a more energy-secure system with a larger proportion of renewable energy resources in the energy mix. This has involved removing subsidies on several non-renewable energy sources and tax incentives for renewable energy technology imports. In this vein, housing cooperatives may be able to link with energy cooperatives to take advantage of this policy directive and the capacity to pool resources across their members to engage in the micro-generation of renewable energy, e.g. household solar panels or wind turbines, at the property level.

Strengthening the link between micro-generation of energy and housing cooperatives seems to be a fit in terms of how they can both help achieve a society goal by providing more reliable access to energy and helping limit GHG emissions. A further role for the cooperative occurs because the upfront expenditure on micro-generation can be quite significant in terms of the purchase and installation of solar panels as well as suitable battery storage to store energy that is not immediately consumed or fed back into the overall grid. This can be beyond the means of individual cooperative members but by pooling resources it may become affordable that properties geographically close to one another are now able to purchase the technology and link their properties to the micro-generation capacity. Moreover, if the cooperative has a diversified portfolio of properties and members it may be able to raise further capital for investment more widely or provide a socially focused subsidy on the properties where this can be taken up.

3 Housing Cooperatives: Climate Change Adaptation Opportunities

3.1 Housing Cooperatives and Adaptive Capacity

While there are several views on climate change adaptation, one is that climate change adaptation should address the root causes of vulnerability where possible (Berrang-Ford et al. 2021). When adapting to climate change, IPCC (2022) indicates three areas that one can work against to limit impacts. This can be hazards (e.g. changes in windstorm extent), exposure (e.g. what can be threatened by climate change) and vulnerability (e.g. how susceptible people are to a negative impact). Altering these three aspects of climate change impact can be relatively simple (e.g. altering working hours to avoid high temperatures to more complex procedures such as flood-proofing their residence whereby people replace materials with more waterproof materials or seal potential water entry points into their property. A question that may readily come to fore is how visible housing cooperatives are in supporting climate change adaptation.

There has been a large amount of research into the behavioural impulse behind why people do or do not adapt to climate change and its resulting impacts, see, e.g. van Valkengoed and Steg (2019). While there are many frameworks and findings, a consistent finding is that housing security is a major factor in a household's ability to protect and recover from climate-related risks and disasters (Carriere and Lesnikowski 2024). This is because we can understand strong property rights can establish ownership of the alterations as well as the legal right to do so (e.g. tenants cannot unilaterally make changes, particularly fundamental changes to structural designs).

The larger percentage of those in possession of houses in urban Nigeria are tenants, hence they lack the legal capacity to alter housing design to suit climate change mitigation. On the other hand, property owners may not be encouraged to redesign their houses to suit climate change mitigation efforts for some visible reasons, the most profound of these reasons being low personal financial return on investment and the lack of legislative compulsion to effect such changes.

Housing cooperatives can interact with climate change adaptation by enabling access to property ownership within the constraints of cooperative ownership. This can enable people to adapt who would otherwise not have the capacity to do so. Moreover, while ownership is shared this can mean that recurring adaptations to climate change or disaster risk can be established as part of the cooperative's covenant for participation. This aligns with cooperative ethics such as Democratic Member

Control; Education, Training and Information; Cooperation among Cooperatives; Concern for Community; Self-help; Self-responsibility; Equity and Solidarity.

Additionally, the constant presence of the cooperative as an institution that survives through the changes in occupancy as people move into and out of the property allows the cooperative to act as an aggregator of experience and institutional memory. The promotion, maintenance and improvement of a source of institutional memory allows for this evolving set of knowledge to be transmitted to the participants of the cooperative. While providing information itself may promote sustained climate change adaptation activities (Dillenardt et al. 2024; Rollason et al. 2018; Snel et al. 2019), having information coming through trusted channels can promote stronger and more reliable actions. The angle that is provided by the cooperative to be a trusted channel of information can also allow for the improvement of other sources of adaptive capacity. For example, trust can be considered as an element of social capital, which can have knocks or implications for peoples' levels of risk perception (to drive the perceived need to act) and peoples' coping capacities (the self-perceived capacity to act) as an individual has social ties that allow the transfer of knowledge or an expectation of aid when required.

3.2 Housing Cooperatives and Financial Aid

Not only is a housing cooperative a mechanism for promoting social capital and wider adaptive capacity, but the observation that a cooperative allows for the pooling of resources can lead to other aspects of climate change adaptive potential to be improved. For example, the further observation that Nigerian cooperatives have a financial aspect to them as well can be further seen as a synergistic element with housing cooperatives. For instance, a housing cooperative that extends past a single building or community can create a geographically diversified portfolio of properties and participants. Diversification is a key element of overall risk management as it means that one asset can be damaged while another is not; this means that only a portion of wealth is lost, and revenue can still be generated from other assets.

As stated earlier, many Nigerian cooperative societies are primarily financial cooperatives. The trend with many Nigerian cooperative societies is to diversify, for example, into transportation business, consumer cooperatives, land acquisition and sharing business and in some cases housing cooperatives. It flows naturally that successful housing cooperatives easily diversify into other forms of businesses such as care for the elderly, children's daycare, sewage management business and more adventurous such as commodity shops, chains of commodity shops, gas stations, etc.

Additionally, the diversification of properties is a key aspect of insurance funds (Charpentier 2008), which can be structured by sojourning part of the management costs into a fund that can be used to compensate for disaster losses or fund repairs. However, the creation of such a fund requires a significant degree of scaling so that a suitably secure and stable provisioning of revenue can be developed without the potential for catastrophic losses to undermine the funds' viability. Therefore, it may take a significant period of development before such a community-focused fund can be created. The creation of an insurance fund via a housing cooperative has a further advantage in that it can be socially rather than commercially orientated. The absence of a commercial orientation could allow premiums to fall compared to what would be commercially available, as there would be a lower need for a profit and cost-loading factor on top of the expected risk (Hudson et al. 2016).

While the cooperative-owned and managed insurance requires the cooperative in effect to operate at a national scale, smaller-scale cooperatives may also be able to leverage more advantageous agreements with insurers. This may allow easier access to insurance coverage and protection at an individual level or via group insurance through parametric insurance products.

4 The Potential Role of Novel Technologies to Support Housing Cooperatives

Novel technologies can significantly enhance the ability of housing cooperatives to support climate change adaptation and disaster risk reduction. This is because they either enable action directly or offer new opportunities to grasp.

The previous sections noted that a suitably focused charter can incentivise climate action; however, if reliable energy and internet provision can be used then their opportunities through the concept of 'Smart Buildings' and the 'Internet of Things (IoT)' based on multiple sensors across the cooperatives properties (Poyyamozhi et al. 2024; Smeenk and Petock 2023). These sensors can be used to monitor a range of property-level environmental conditions such as temperature, humidity, air quality and structural integrity for early warning of risks like overheating or mould. This allows for increased energy resilience by allowing for more automated or rigorous control over environmental conditions, which can prevent energy wastage or further health issues (e.g. mould outbreaks worsen ill health via breathing conditions). When coupled with a charter that promotes the use or (retrofitting of) energy-efficient building materials this can create a double dividend in reducing GHG emissions from a property portfolio. Additionally, emerging technologies such as 3D printing may be able to offer innovations in building materials and processes that were inaccessible before (Sakin and Kiroglu 2017). A housing cooperative may be able to take advantage of such novel materials and production processes because the absence of a strong commercial profit motive combined with a motive towards generating social value can allow for more innovation and risk-taking in this area. For example, this may allow them to explore localised, cooperative-owned biological treatment systems or greywater recycling to reduce water dependency through improved recycling capacities.

As currently operated, Nigerian cooperatives have not fully aligned with the use of novel technologies with some cooperative societies showing very low levels of usage. Invariably, the foundation upon which harnessing the potential inherent novel technologies may be built is currently fraught with low patronage; hence, utilisation will be sub-optimal. Addressing the situation will require a departure from current practice.

Another emerging technology that may prove useful in linking housing cooperatives to improved climate resilience is via more secure property rights, which are publicly validated transaction records in a public blockchain ledger (Konashevych 2020). A public blockchain ledger has the advantage of creating a secure record of ownership and property rights which creates an increasingly secure tenure of ownership, which has been trialled or proposed in several countries across the globe (Mintah et al. 2021; Yapicioglu and Leshinsky 2020). Insecure property rights are a significant hurdle to overcome in many regions of the Global South, unless this is overcome, we do not have the capacity to maximise property owners or residents to adapt to climate change. The availability of secure and reliable property rights is important because adapting to change patterns of climate and disaster risk is a long-term process. This is because actions taken today may entail a rather large upfront expenditure and accrue benefits slowly over a long period of time, e.g. property-level flood protection is cost-effective but may have a 15–20-year payback period that is often through a lower intangible level of risk (Hudson et al. 2014) rather than a direct monetary benefit. Therefore, secure tenure is fundamental so that a person or organisation believes that they will be able to capture these benefits directly or indirectly. A further aspect of blockchains that may be beneficial is that it can be argued to enable more transparent governance of the collective, for example through a more transparent and recordable distribution of funds within and across the collective. Secured property rights through public blockchain ledger and allied initiatives require amongst others, supportive legal frameworks. This contrasts with the current state of Nigerian cooperative law which remains largely a relic of the Nigerian Cooperative Ordinance of 1935, significantly unsuitable for modern trends in cooperatives and housing cooperatives.

5 Limitations Faced by Housing Cooperatives for Proactive Social and Environmental Action

While there are several potential benefits that housing cooperatives provide, there are also limitations that they face. The first is simply the financial constraints that they face with their current size, capitalisation and potential access to further investment or financing (Chiquier and Lea 2009; International Cooperative Alliance 2017). Cooperatives often face barriers in accessing climate finance, government grants, or private sector investments due to their non-profit or community-led nature which leads them to be considered a relatively higher risk compared to other opportunities. This is a significant barrier because not only are their high upfront costs for implementing climate-resilient buildings (e.g. energy-efficient retrofits, flood-proofing), that are potentially more expensive for retrofits rather than construction but also land prices and construction costs in many urban areas of Nigeria can be beyond the means of the many cooperatives to afford in any meaningful way. This limitation has the additional negative consequence that it results in potential cooperative investors acquiring former farmland for conversion into housing stock. This corresponds to increasing land use land cover change and environmental loss in Nigeria. The environmental loss that this represents can be considered counterproductive because the conversion of land use releases carbon on top of the further loss of potential natural carbon sinks. The urban sprawl that seeks cheaper land for conversion or purchase can further worsen urban wellbeing because residences are placed further away from key infrastructure (e.g. roads, energy networks, healthcare), which undermines several of the Sustainable Development Goals. This in turn can be worsened by the often absence of strict planning regulations and monitoring in the Global South, which can result in urban development that is not integrated with wider urban planning concerns.

A related limitation can be common models of cooperative governance and decision-making. For example, the primary driver of proactive environmental action will be driven by the overall collective will of the cooperative members. Therefore, we cannot argue that all cooperatives will make the same choice regarding the trade-off between the affordability of climate action and may decide that it is not in their best interest to act in a sustainable manner at any point in time. Similarly, depending on how democratic the coverage of the cooperative wishes to be, we can see that there is slower decision-making as key actors within the cooperative's governance structure try to develop suitable consensus-based actions which can delay the adoption of adaptation and mitigation strategies. This problem can be worsened if there are diverse interests and priorities within the cooperative's membership, meaning that a consensus is harder to develop. While this collective action problem can be managed by selectively allowing only certain individuals into the group to maintain a cohesive vision of the cooperative, this may become discriminatory if left unchecked.

These issues may be further compounded by a lack of supportive policies. This means that Nigeria may not have established a suitable regulatory framework that supports climate-adaptive cooperative housing in a sustainable manner. This can therefore create difficulty in scaling up suitable models of cooperative housing limiting the development of cooperative housing to small and midscale operations, which, for example, would prevent suitable access to a diversified portfolio of properties and members to base a communal insurance fund upon.

6 Conclusion and Recommendations

Housing cooperatives present a promising mechanism for addressing both climate change mitigation and adaptation, particularly in rapidly urbanizing countries like Nigeria. By embedding principles of shared ownership, social responsibility and long-term sustainability, housing cooperatives have the potential to enhance environmental governance at, effectively, the community level. Their ability to self-regulate building practices, encourage behavioural change and promote energy efficiency underscores their potential role in reducing greenhouse gas emissions while simultaneously improving social equity and housing affordability.

Furthermore, housing cooperatives can bolster climate adaptation by strengthening institutional memory, fostering social capital and providing financial mechanisms for resilience-building. Their cooperative structures can facilitate pooled financial resources for climate-resilient infrastructure and recovery financing, while their potential linkages with energy cooperatives may further enhance access to sustainable energy solutions. Emerging technologies, such as blockchain for secure property rights and smart monitoring systems, could further support the viability and effectiveness of housing cooperatives in climate action.

However, significant challenges remain. Access to finance, policy support and governance structures that balance democratic decision-making with efficient action are critical hurdles. Without supportive legal frameworks, incentives and capacity-building efforts, housing cooperatives may struggle to scale their impact. Additionally, land acquisition challenges and potential conflicts between collective ownership and individual interests must be carefully managed to avoid unintended negative consequences. The most significant challenges to overcome will be the creation of a suitable enabling regulatory framework that enables housing cooperatives to develop and reach a significant scale where they can be a viable, systematic, alternative to the current conceptualisation of housing provision in Nigeria.

Currently, viable models of housing cooperatives are replete in some countries of the global north. These models may be adapted as supplement to homegrown initiatives of housing cooperatives to the extent that local peculiarities permit to create a hybrid model of resilient Nigerian housing cooperatives. The foregoing may serve as templates in other countries of the Global South.

Declaration of Competing Interests There are no competing interests to report.

References

Abiodun O, Salami T (2023) Land use dynamics of urban change in the Lagos-Ogun urban fringe, Southwest Nigeria. Urban Regional Plan Rev 9(1):78–88

Akanji AA (2022) Cooperative societies and the sustainable development goals (SDGs): a viable model for human development in Nigeria. Polit Econ Colon Nation-Build Niger:345–360

Balmer I, Gerber J-D (2018) Why are housing cooperatives successful? Insights from Swiss affordable housing policy. Hous Stud 33(3):361–385

Berrang-Ford, L., Siders, A. R., Lesnikowski, A., Fischer, A. P., Callaghan, M. W., Haddaway, N. R., Mach, K. J., Araos, M., Shah, M. A. R., Wannewitz, M., Doshi, D., Leiter, T., Matavel, C., Musah-Surugu, J. I., Wong-Parodi, G., Antwi-Agyei, P., Ajibade, I., Chauhan, N., Kakenmaster, W. et al (2021). A systematic global stocktake of evidence on human adaptation to climate change. Nat Clim Chang, 11(11), 989–1000. doi:https://doi.org/10.1038/s41558-021-01170-y

Boateng GO, Balogun MR, Dada FO, Armah FA (2020) Household energy insecurity: dimensions and consequences for women, infants and children in low- and middle-income countries. Soc Sci Med 258:113068. https://doi.org/10.1016/j.socscimed.2020.113068

Carriere ZD, Lesnikowski A (2024) The climate change adaptation readiness of co-operative housing in Nova Scotia Canada. Environ Policy Governance 34(6):724–739. https://doi.org/10.1002/eet.2116

Cabeza LF, Vérez D, Zsembinszki G, Borri E, Prieto C (2022) Key Challenges for High Temperature Thermal Energy Storage in Concrete—First Steps towards a Novel Storage Design. Energies 15(13):4544. https://doi.org/10.3390/en15134544

Charpentier A (2008) Insurability of climate risks. Geneva Pap Risk Insur-Issues Pract 33(1):91–109. https://doi.org/10.1057/palgrave.gpp.2510155

Chilokwu O, Nwakoby NP, Egor EI (2019) Savings diversification for economic growth and development in Nigeria: the co-operative model. J Glob Account 6(2):79–101

Chiquier L, Lea M (2009) In: Chiquier L, Lea M (eds) Housing finance policy in emerging markets. World Bank Publications. https://www.mfw4a.org/sites/default/files/resources/Housing%20Finance%20Policy%20in%20Emerging%20Markets.pdf

Co-operative Housing Federation of Canada (2025) Environmental sustainability and co-ops. https://chfcanada.coop/about-co-op-housing/environmental-sustainability-and-co-ops/

Crown Commercial Service (2025) Social value in procurement—procurement essentials. https://www.crowncommercial.gov.uk/news/social-value-procurement-ccs

Dillenardt L, Bubeck P, Hudson P, Wutzler B, Thieken AH (2024) Property-level adaptation to pluvial flooding: an analysis of individual behaviour and risk communication material. Mitig Adapt Strateg Glob Chang 29(6):53. https://doi.org/10.1007/s11027-024-10148-y

El-Bouayady R, Radoine H (2023) Urbanization and sustainable urban infrastructure development in Africa. Environ Ecol Res 11(2):385–391

Gago EJ, Roldan J, Pacheco-Torres R, Ordóñez J (2013) The city and urban heat islands: a review of strategies to mitigate adverse effects. Renew Sust Energ Rev 25:749–758

Ganapati S (2014) Housing cooperatives in the developing world. In: Affordable housing in the urban global south. Routledge, pp 102–116

Hudson P, Botzen W, Kreibich H, Bubeck P, Aerts J (2014) Evaluating the effectiveness of flood damage mitigation measures by the application of propensity score matching. Nat Hazards Earth Syst Sci 14(7):1731–1747

Hudson P, Botzen WJW, Feyen L, Aerts JCJH (2016) Incentivising flood risk adaptation through risk based insurance premiums: trade-offs between affordability and risk reduction. Ecol Econ 125:1–13. https://doi.org/10.1016/j.ecolecon.2016.01.015

Hufen J, Koppenjan J (2015) Local renewable energy cooperatives: revolution in disguise? Energy Sustain Soc 5:1–14

International Co-operative Alliance (2017) The capital conundrum for co-operatives. https://ica.coop/sites/default/files/publication-files/ica-the-capital-conundrum-for-co-operatives-en-1892359719.pdf

International Co-Operative Alliance (2025) What is a cooperative? https://ica.coop/en/cooperatives/what-is-a-cooperative

IPCC (2022). Climate change 2022: impacts, adaptation, and vulnerability. In: Pörtner HO, Roberts DC, Tignor M, Poloczanska ES, Mintenbeck K, Alegría A, Craig M, Langsdorf S, Löschke S, Möller V, Okem A, Rama BM (eds) Contribution of working group II to the sixth assessment report of the intergovernmental panel on climate change. Cambridge University Press

IPCC I (2007) Climate change 2007 mitigation of climate change. IPCC

Konashevych O (2020) Constraints and benefits of the blockchain use for real estate and property rights. J Property, Plan Environ Law 12(2):109–127

Lillo P, Ferrer-Martí L, Juanpera M (2021) Strengthening the sustainability of rural electrification projects: renewable energy, management models and energy transitions in Peru, Ecuador and Bolivia. Energy Res Soc Sci 80:102222

Marckmann B, Gram-Hanssen K, Christensen TH (2012) Sustainable living and co-housing: evidence from a case study of eco-villages. Built Environ 38(3):413–429

Mintah K, Boateng FG, Baako KT, Gaisie E, Otchere GK (2021) Blockchain on stool land acquisition: lessons from Ghana for strengthening land tenure security other than titling. Land Use Policy 109:105635

Monyei CG, Akpeji KO, Oladeji O, Babatunde OM, Aholu OC, Adegoke D, Imafidon JO (2022) Regional cooperation for mitigating energy poverty in sub-Saharan Africa: a context-based approach through the tripartite lenses of access, sufficiency, and mobility. Renew Sust Energ Rev 159:112209

Newell P, Daley F, Twena M (2022) Changing our ways: behaviour change and the climate crisis. Cambridge University Press

Noah IA (2024) Nigerias climate change act, conditional and unconditional nationally determined contributions, and the principle of common but differentiated responsibilities. Environ Law Rev 14614529241272183. https://doi.org/10.1177/14614529241272183

Onanuga MY, Eludoyin AO, Ofoezie IE (2022) Urbanization and its effects on land and water resources in Ijebuland, southwestern Nigeria. Environ Dev Sustain 24(1):592–616

Ostrom E (1990) Governing the commons: the evolution of institutions for collective action. Cambridge university press

Pistocchi A (2017) Hydrological impact of soil sealing and urban land take. In: Urban expansion, land cover and soil ecosystem services. Routledge, pp 157–168

Pokubo D, Al-Habaibeh A (2023) Study of the underlying factors driving household renewable energy transitions in Nigeria. WIT Trans Ecol Environ 261:115–127

Poyyamozhi M, Murugesan B, Rajamanickam N, Shorfuzzaman M, Aboelmagd Y (2024) IoT—A promising solution to energy Management in Smart Buildings: a systematic review, applications, barriers, and future scope. Buildings 14(11):3446

Punt MB, Bauwens T, Frenken K, Holstenkamp L (2022) Institutional relatedness and the emergence of renewable energy cooperatives in German districts. Reg Stud 56(4):548–562

Rollason E, Bracken LJ, Hardy R, Large A (2018) Rethinking flood risk communication. Nat Hazards 92:1665–1686

Sacchetti S, Tortia EC (2014) The social value of multi-stakeholder cooperatives: the case of the CEFF system in Italy. In: Research handbook on sustainable co-operative enterprise. Edward Elgar Publishing, pp 285–300

Sakin M, Kiroglu YC (2017) 3D printing of buildings: construction of the sustainable houses of the future by BIM. Energy Procedia 134:702–711

Schipper EL, Ayers J, Reid H, Huq S, Rahman A (2014) Community-based adaptation to climate change: scaling it up. Routledge

Schmid AA (2008) Conflict and cooperation: institutional and behavioral economics. John Wiley and Sons

Smeenk HG, Petock M (2023) Internet of things for smart buildings: leverage IoT for smarter insights for buildings in the new and built environments. Packt Publishing Ltd.

Snel KA, Witte PA, Hartmann T, Geertman SC (2019) More than a one-size-fits-all approach–tailoring flood risk communication to plural residents' perspectives. Water Int 44(5):554–570

Sun J, Weng L (2024) Dynamics of mitigation measures in urbanizing Africa: challenges and opportunities for future climate change. J Clean Prod 141443

United Nations (2022) Population. https://www.un.org/en/global-issues/population

United Nations Department of Economic and Social Affairs (2018) 68% of the world population projected to live in urban areas by 2050. https://www.un.org/uk/desa/68-world-population-projected-live-urban-areas-2050-says-un

van Valkengoed AM, Steg L (2019) Meta-analyses of factors motivating climate change adaptation behaviour. Nat Clim Chang 9(2):158–163. https://doi.org/10.1038/s41558-018-0371-y

Yapicioglu B, Leshinsky R (2020) Blockchain as a tool for land rights: ownership of land in Cyprus. J Property, Plan Environ Law 12(2):171–182

Zelizer VA (2012) How I became a relational economic sociologist and what does that mean? Polit Soc 40(2):145–174

NIMBY to YIMBY: Towards a Better Neighbourhood

Li Li, Jingyuan Ren, and Blessings Chiwosi

1 Introduction

Municipal solid waste (MSW) management related to methane (CH_4)[1] emission in landfills and their curbing measures is a topic worth exploring in low-carbon governance practices. As a strategy designed to address social issues arising from improper solid waste disposal (Henry et al. 2006; Nemerow et al. 2009; Wilson 2007), MSW is a specific practice with different features in various countries and communities (Nathanson 2015).

Drivers might be found in public health, resource scarcity, value of waste, public awareness, participation and so on (Marshall and Farahbakhsh 2013). Though using landfill as a method for managing solid waste requires high site standards and incurs high costs due to the large land area it utilises (Das et al. 2019), using high-temperature oxidation through waste incineration reduces the volume of waste by 70–80%, and effective landfilling or further processing after incineration can reduce land resource usage by about 90%, thereby ameliorating land resource efficiency (Cheng and Hu 2010; Zhang et al. 2015). In addition, waste incineration can be a more advanced and environmentally friendly method than landfill (Bosmans et al. 2013). The grate combustion process used in waste incineration power generation provides better pollution control, faster processing and greater volume reduction (Arena 2012),

[1] Methane is responsible for around 30 per cent of global warming since the Industrial Revolution. It is a potent greenhouse gas with a global warming potential much higher than carbon dioxide (CO_2) (about 28–36 times more potent over a 100-year period).

L. Li (✉) · B. Chiwosi
College of Humanities and Development Studies/College of International Development and Global Agriculture, China, Agricultural University, Haidian District, Beijing, China
e-mail: lilian8006@cau.edu.cn

J. Ren
Institute of International and Regional Studies, Southeast University, Nanjing, China

M. Okitasari et al. (eds.), *Equitable and Sustainable Cities*, Advances in 21st Century Human Settlements, https://doi.org/10.1007/978-981-92-2793-8_5

extending the lifespan of landfills and producing heat for power generation (Assamoi and Lawryshyn, 2012).

However, opponents argue that both methods of disposing of solid waste are problematic (Kasperson et al. 1992). Studies point out that landfill waste treatment technologies are still hazardous and can produce a large amount of leachate (Schiopu and Gavrilescu 2010), which can infiltrate through the gaps in the protective films, causing groundwater pollution (Njoku et al. 2019). Besides, landfills are associated with reproductive risks, such as increased incidence of congenital malformations (Fielder et al. 2000).

Scholars also study residents' perceptions of the impact of landfill establishment. One study conducted in Palestine estimates that their preferences might be most negatively influenced by air pollution, land value loss and leachate leaks (Salah et al. 2020). On the living front, the smell of landfill leachate can be a cause of public discontent (Che et al. 2013; Ozbay et al. 2021). Those who oppose the construction of waste incineration power plants claim they emit carcinogenic substances, including dioxins, mercury and furans (McKay 2002). A study even found that there are anomalies and lower birth weights in newborns near incineration plants (Mattiello et al. 2013). Emissions from incineration plants are associated with increased respiratory symptoms, such as coughing, wheezing, asthma and even an increased risk of lung cancer (Porta 2009).

Under the self-interest model of attitudes, few urban residents would like to face the potential risks of new MSW technologies. The phenomenon of 'not in my backyard' (NIMBY), or even that of 'not in anybody's backyard' (NIABY), might be related to the hardship of accurately predicting negative effects and balancing financial compensation with an inner sense of inequity (Lober and Green 1994). Since the 1980s, 4230 cases have been reported by the Global Environmental Justice Atlas (EJAtlas) on various environment-related conflicts (Temper et al. 2015; EJAtlas 2025), showing the challenges faced by localities.

The siting of waste incineration has not only been a highly contentious issue in Western countries by means of lobbying, litigation and extra-institutional tactics (Rootes and Leonard 2010; Walsh et al. 1997) but also an extremely pressing issue in China. Top-down and non-consultative approaches led by governments have encountered social opposition, such as the three anti-incineration campaigns in three districts of Beijing (Johnson 2013).[2] However, instead of being fought against, the waste incineration project in Anding Town of Daxing District has been welcomed. It is important to explore what took place during this peaceful relocation progress so as to dig deeper into how disputes have been minimised or even derived into an effective governance pattern.

[2] Locations of the three places are at Liulitun Village (Shangzhuang Town, Haidian District), Gao'antun Village (Changying Town, Chaoyang District), and Asuwei Village (Xiaotangshan Town, Changping District).

2 From NIMBY to YIMBY: What Matters?

Referring to technological applications from landfill to waste-to-energy (WTE) incineration, both public policies and private incentives are important in generating No-to-Yes enablers for public engagement and inner motivations. According to current literature studying MSW-related technologies, the roots for positive changes have been studied from the following aspects.

2.1 Bottleneck in Recycling Capacity Limits

Globally, about 1.5–2.1 billion metric tonnes of municipal waste have been generated annually since the 1950s in China. It means about 400 kg of waste is discarded by urban citizens every day (Yu and Qin 2005). In developed countries like Finland or Switzerland, waste classification and treatment are better done, and the recycling rate of waste exceeds 60%. However, in developing countries, only about 16% of the waste could be recycled, with most discarded without proper reuse or eliminated unsustainably, causing air, soil or water pollution (Stoica 2022). With the rapid increase in municipal waste volume, many landfill sites have to retire earlier than expected (Wang 2020). However, landfill site expansion is always protested (Sharholy et al. 2008; D'Alisa et al. 2010; Dines 2013; Tomé 2023; Jiang 2023).

The bottom-up conflicts between community residents and local government organisations in different places show 'ecological modernisation' towards the inconvenient truth (Lang and Xu 2013), with opponents pushing the government to cancel or adjust the siting to less informed residents, as seen in the cases in Liulitun and Asuwei in Beijing. For the Liulitun project, the government changed the site to another town in Sujiatuo in the Haidian District. The Asuwei project in Xiaotangshan Town of Changping District was halted for 5 years and restarted as a comprehensive treatment site for both landfill and incineration in 2017, with residents resettled (Collaborator 2021; Demaria 2021). Technological progress of incineration is tackling the bottleneck of recycling capacity to build more sustainable cities in the process of urbanisation, but the scaling-up is still faced with political challenges.

2.2 Situational Legitimacy of Local Governments

A quantitative analysis of 10 cases of waste incineration or paraxylene (PX) plants in different provincial urban regions has been carried out to see under what conditions the local governments of China compromise with NIMBY activists. It is identified that the absence of support from the national government is a necessary condition for local governments to change their former decisions (Li et al. 2016).

This recalls the 'situational legitimacy' calibration of social organisations in Chinese local communities. With overall legitimacy under specific projects from the central government, social organisations in China need to be recognised by both their customers and grassroots elites in local communities (Deng 2019).

However, the key issues are not about how the government compromises but how to engage residents in an imminent process with their own intentional potential to achieve modernity and prosperity through effective measures as solutions towards a sustainable future.

2.3 Multiple Decision-Makers for the Backyards

Orchestration of MSW involves stakeholders with distinct roles and responsibilities. Those who are typically governments, urban planners, corporations, NGOs or individuals responsible for specific MSW functions or aspects (Joseph 2006).

Awareness and participation of public stakeholders are important drivers of MSW (Marshall and Farahbakhsh 2013). The concept of NIMBY was first proposed by Michael O'Hare (O'Hare 1977; Dear 1992), and later so-called NIMBY facilities indicate the non-inviting actions with fear of lowering life quality or property value of residents (Chung et al. 2016; Vittes et al. 1993). While protests seem hard to avoid (Huang and Yang 2020), some believe that NIMBY conflicts can be addressed, as long as risk management mechanisms with participatory principles and models are created (Xu and Jin 2022; Van den Akker 1999; Nelson and Wright 1995; Fischer 2012, 2000; Stringer and Reed 2007).

Street-level bureaucrats, as the actual implementers of public policies with discretion, play double-edged roles in balancing public and private needs (Lipsky 1980). However, in the green transition of the Chinese mainland, 'grassroots cadres (*jiceng ganbu*)' provide an aspect of economic enforcement recipes as situational innovative solutions (Strauss 2009).

Private empowerment is another driver. Bottom-up participatory governance models are set up, prioritising the participation of citizens in decision-making, enabling them to have a direct impact on policymaking (Lidskog 1997; Devas and Grant 2003; Nikkhah and Redzuan 2009; Tian and Han 2022; Jiang 2023; Lu 2023). Enterprises could also be enabled to calibrate synergies dynamically through E-governments and information and communication technology (ICT), converting the formerly vertical relations into horizontal ones with the facilitating roles played by social media (Young 2013; Scupola and Zanfei 2016; Li and Ding 2020). It is not only about providing resources or opportunities but also about giving a voice to the target groups, especially the vulnerable (Turner and Maschi 2015), developing the capacity and confidence of individuals and communities to act and make choices (Schuftan 1996).

2.4 *Holistic Governance Calibrating Synergies Needed*

Based on the above aspects, synergies need to be calibrated holistically to achieve equilibrium between growth and degrowth (Hickel et al. 2022). Holistic governance is definitely access to consolidate fragmented actors towards solidarity, especially in cross-sectors of environmental protection (Li and Ding 2020: 3; Rummery 2006; Li et al. 2018). Different from forceful powers guaranteeing social order (Hobbes 1651; Goodnow 1900), or rationality of Homo economicus maximising benefits in a market-based manner (Buchanan and Tullock, 1960; Hayek 1944), holistic governance believers hold that mass needs are not given a priori. When collective good is obviously an obligation, mass willingness can be constructed towards better provision. Public sectors could indirectly manipulate 'pseudo or quasi-markets (Le Grand 1991)' like carbon trading markets, or harness public–private balance through 'calibration' (Dunsire 1990).

It is significant to summarise experience from practically realised MSW governance cases, such as the largest-scale one in Anding Town. Field research and semi-structured interviews with multiple stakeholders have been carried out in this research with the aim of observing why and how consensus has been reached and synergies calibrated.

3 A Prototype Derived from Landfill Practices

In 1997, *the Kyoto Protocol* was reached as the first binding agreement in human history to restrict greenhouse gas (GHG) emissions. It was put into effect in 2005 with 55 national governments' approval, with their GHG emission occupying 55% of the global total. For the first commitment period (2008–2012), the Annex I industrial countries were required to reduce emissions by an average of 5% below 1990 levels. With three flexible mechanisms of Joint Implementation (JI), International Emissions Trading (ET) and Clean Development Mechanism (CDM), a quasi-market mechanism of carbon trading was utilised as a buffer for green transition with priced and exchangeable 'Certified Emission Reductions (CERs)'. Among them, CDM allows Annex I countries to purchase carbon credits from the developing countries in the Global South to fulfil their climate mitigation obligations under the UNFCCC framework, providing a win–win partnering opportunity in between, enabling 'voluntary hybridity' through 'trial-and-error-testing' experiments (Li 2020) in China.

3.1 *The First CDM Project: A Prototype Generator*

The Landfill Gas (LFG) Project of Anding Town was the first trial to apply for the CDM project from China in 2005, with a domestic No.001 certificate approved by

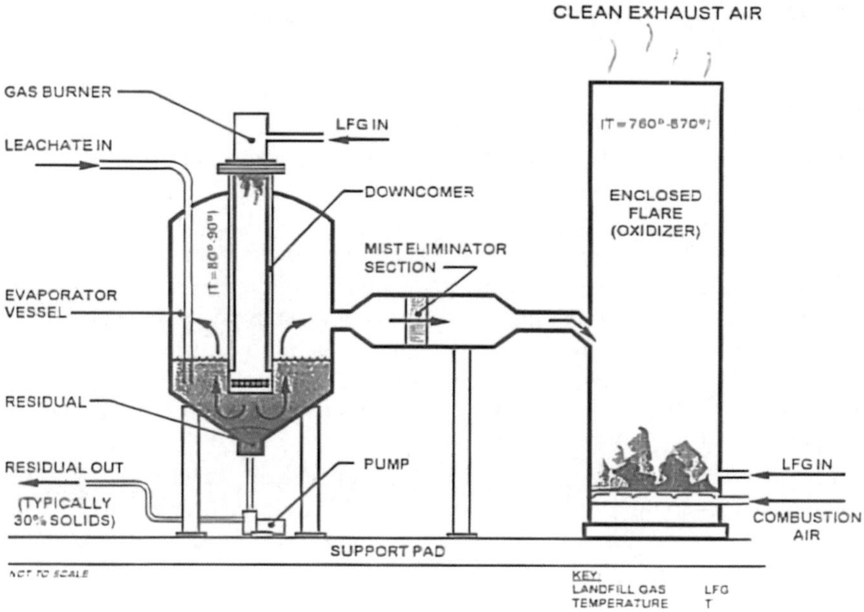

Fig. 1 Schematic of landfill leachate evaporator system (2005–2015). *Source* UNFCCC, Anding Landfill Gas Recovery and Utilisation Project, Project Design Document Form-Version 02, CDM Board, 9 December 2005, Available at: https://cdm.unfccc.int/UserManagement/FileStorage/ILQRLSMFGPEC6PUH53QK5756RAL51U#page=1.23, accessed on 10 January 2025

the National Development and Reform Committee (NDRC). As a landfill site located 40 km South of Beijing, Anding Landfill Site receives about 350,000 tonnes of waste per year (958 tonnes/day). As is shown in Fig. 1, by burning methane (CH_4) into carbon dioxide (CO_2), the global warming potential (GWP) could be reduced largely by a proportion from 1/27 to 1/87 (EPA 2024; IEA 2021). The Dutch company of Energy Systems International (ESI), as the carbon source, purchased the proposed volumes of carbon sink from Beijing Erqing Environment Engineering Group Co. Ltd. (BEEEG) under the CDM framework, starting from July 2002 (Zhang 2007).

3.2 *A Prototype of 'Niche-Nexus-Nodes' Synergy*

Synergy refers to the interaction or creation of a whole of nonlinear addition of forces, energies or effects that are greater than the simple sum of its parts, which originates from an Attic Greek word meaning 'working together' (Jaffe 2010). As represented in Table 1, BEEEG created synergistic effects with three groups of partners to achieve three kinds of 'whole':

Table 1 'Niche-Nexus-Nodes' synergy in No.001 CDM LHG project in China

Synergy created	Actors	Levels	Ownership	Needs	Provision
Niche	BEEEG	Anding Town	SOE (Hybrid)	MSW dealing technology to offset 75,557 tCO_2e/year; 755,568 tCO_2e in total	Carbon sink
	ESI (the Netherlands)	Transnational	Private	Emission reduction from 4,483–5,251 tCO_2e (2004) down to 1,551 tCO_2e (2014)	Carbo source
	TPI (New Zealand)	Transnational	Private	International businesses	Percolate tech-transfer
Nexus	UNFCCC/ CDM	Global	Public	*Kyoto Protocol* Implemented in Non-Annex I countries	CDM platform
	NDRC/ MOST/ MFA/ Beijing	Ministerial/ Provincial	Public	First CDM project/ Pilot: National Threshold Price (USD 5 in 2004 & EUR 8 later)	*Kyoto Protocol* implementing
Nodes	Daxing District	Local	Public	Fewer Public Protests	MSW services
	Beijing Jifeng Industrial Investment Management Co. Ltd. (HK)	Transnational	Private	CDM expertise: merging into ESI; active GHG credits buyer for LFG, wind power, hydropower, etc	20 million RMN yuan

Source UNFCCC (2024). "Project 0296: Anding Landfill Gas Recovery and Utilisation Project", 24 October, https://cdm.unfccc.int/Projects/DB/DNV-CUK1141654012.19/view, accessed on 24 October 2024. Li Li (2020). "'Green' effects of hybrid actors through carbon trading: Cases in Beijing", Global Transitions Proceedings, 1, 16–17.

- Niche of technologies or pilot policy zones.
- Nexus among partners or stakeholders.
- Nodes in issue-linked networks.

Later, with provincial CDM Centres established, providing consultancy services and enabling enterprises to join the international carbon trading of UNFCCC, China became the largest CDM host country in 2006, with 3863 CDM projects applied, ranking the top (UNFCCC 2024). Innovative capacities in renewable energy (RE) technologies were also built, and talents were nurtured during such a public–private mutually empowering process (see Fig. 2).

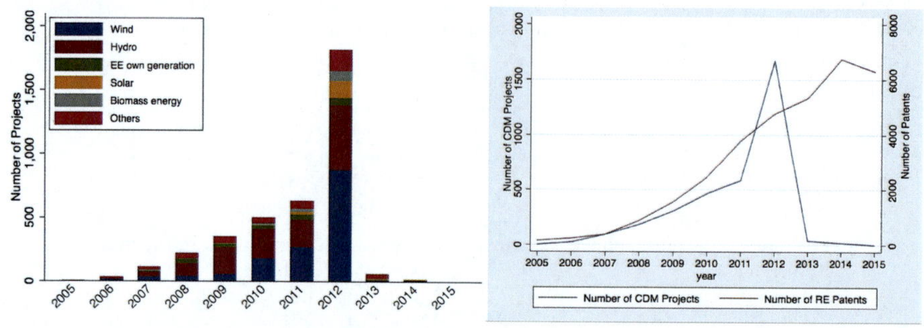

Fig. 2 Number of CDM projects sectors & RE Patents in China (2005–2015). 2020 *Source* Cui et al. (). Can CDM projects trigger host countries' innovation in renewable energy? Evidence of a firm-level dataset from China. Energy Policy, 139, 1–10

4 Prototype Application to MSW Incineration

From 1979 to 2003, the capacity of MSW treatment was consolidated and enhanced dramatically (see Fig. 3). Regulated in 1988, the earliest landfill criteria in China could guarantee a good order of MSW siting, construction and management, but expansion of landfill sites has met a bottleneck. The first MSW incineration site was also set up in Shenzhen, Guangdong Province, in southern China. Introducing Japanese technology as a first move, its daily waste-treating capacity was only 100 tonnes at that time, which was a great achievement (Wang et al. 2015). Since the late 1980s, innovative approaches to building MSW sites into beautiful and eco-friendly parks emerged, such as the Kasai Rinkai Park in Japan (Li and Ding 2016).

NIMBY attitudes started to change with both visible and value-added discussions about cleaner and more acceptable choices in the early twenty-first century in China. Compared with landfill technologies with a price of 6–11 USD/tCO_2e (40–80 RMB yuan) or a monetary benefit of 205–1260 USD/day (1500–9200 RMB yuan/day), incineration is much more lucrative for about 39 million USD/day (280 million CNY/day), which is tens of thousands of times more than the net earnings of landfill percolating (Zhao et al. 2016). Since China's 11th Five-Year Plan (2006–2010), waste incineration was encouraged for the first time in public policies. By 2010, in a total of over 680 municipalities in China, 1/4 lacked new proper sites for landfills.

Considering this, the '12th Five-Year Plan (2010–2015)' of national urban waste disposal facilities set a target of increasing the incineration proportion of the primary energy from 13% in 2010 up to 45% in 2015, lowering landfill from 73% down to 30% (The State Council 2012). Shenzhen, a metropolis with extremely limited land, proposed an ambitious goal of 100% waste incineration, together with incentive policies in tax and electricity price subsidies to boost waste incineration (Lu 2023). In 2013, Lujiashan incineration site, by then the world's largest WTE plant in a single phase was constructed in Mentougou District of Beijing (Gao 2016). In 2015, a state-owned enterprise (SOE), Shenzhen Energy Environmental Protection Co., Ltd., proposed the concept of 'creating a park while building a plant', changing the inherent

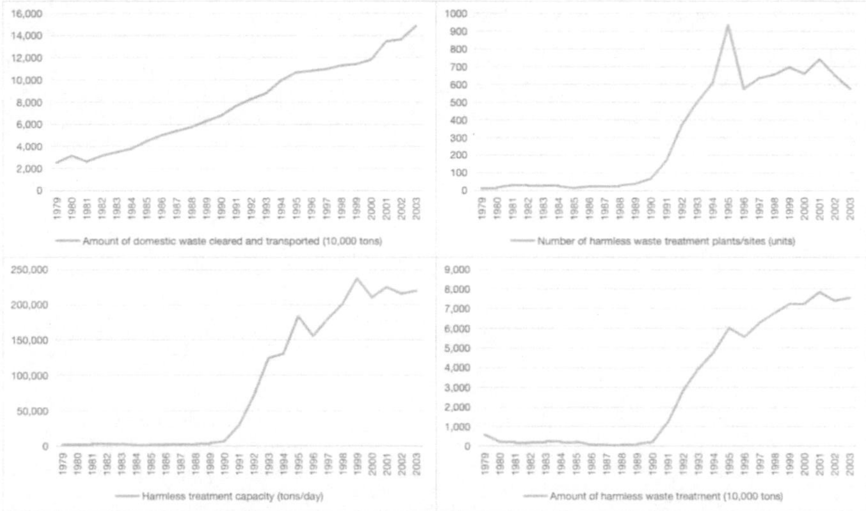

Fig. 3 Harmlessly Treated MSW Statistics in China (1979–2003). Note: From 1980 to 1995, the amount of harmlessly treated waste included both garbage and faeces. 2005 *Source* Yu et al. (). The Experiment of Public–Private Partnerships in China. Shanghai: Shanghai Century Publishing (Group) Co., Ltd. (in Chinese): 207

impression of WTE plants from NIMBY facilities to YIMBY ones (Zhongyuan News 2024). By 2019, 428 incineration sites had been built in China, with another 216 under construction (Wang 2020). Still, by 2020, Chinese MSW amounted to 7 billion tonnes, engulfing over 500 million m^3 of land, costing a loss of about 30 billion CNY. With that pressure, during the '13th Five-Year Plan (2016–2020)', a target of increasing the incineration treatment rate to 50% was proposed (Wang 2020). As the city with the largest MSW dealing pressure, social synergy is greatly needed in Beijing to relieve the unwillingness of residents. The prototype of 'niche-nexus-nodes' synergy is resorted to again to meet the unprecedented requirements of the residents at seven villages in Anding Town[3] to be relocated.

4.1 *Local Voices Heard and Co-designed*

As a suburban district in Beijing with 1.7 million citizens living in 14 towns and six sub-district offices, Daxing is a southern linkage with Tianjin Municipality

[3] These seven villages are relocated to two villa community zones. In the western zone, or "West Garden (*xiyuan*) of Yulin", there are 4 villages, including Zhanshang, Gaodian, Shantaizi, and Zhengfuzhuang. In the eastern zone, or "East Garden (*dongyuan*) of Yulin", there are three villages, including Qianyechang, Houyechang, and Tongzhoumafang. *East Garden of Yulin* is close to an ancient royal mulberry garden, which is unique in Beijing and the largest one of its kind in northern China.

Table 2 Interviewee representatives of Zhanshang Village (2021–2024)

Group Codes	Gender	Age	Identities	Characteristics
M2801	M	28	Village cadre[4]	Actively involved in organising and coordinating the villa building and community affairs
M4702	M	47	Migrant worker	Unique insights into urbanisation
M7303	M	73	Senior farmer	Extensive agricultural experience, living in Zhanshang village for his entire life and knowing its history well
F2404	F	24	Young mother	Unique insights into child education and family life in the village
F3905	F	39	Middle-aged mother	Concerned about ways to improve community welfare and villagers' quality of life
F6706	F	67	Village council member	Profound understanding of village life

Source Authors based on field interviews

and Hebei Province. Integrating a strategy for Beijing-Tianjin-Hebei (*jing-jin-ji*) regional development is taken as a cure for the 'metropolis syndrome' of the capital city of Beijing. Anding Town, with 33 villages, an area of 78 km^2, and a population of over 30,000, is located on the outskirts where villagers possess farmland in horticulture or crops. Since the establishment of the Anding MSW factory in 1996, a 45-m-high garbage pile has gradually formed, covering an area of 16.7 ha. It extends 300 m north to south and 550 m east to west. Among the seven villages, Zhanshang is the closest to the landfill site, only 500 m away.

As is shown in Table 2, semi-structured interviewees are categorised into 6 groups with representatives' opinions integrated into five main themes. All have moved to new venues, at different ages with various backgrounds, half female and half male.

Theme 1: Improvement of public facilities

Villagers between the ages of 20 and 50 expect to live in a place where public amenities could be improved. Otherwise, they intend to leave.

> "In those days the roads outside in the city were wide and smooth, but those in our village were potholed and no one came to be able to fix them."
>
> "There are limited transport for children to go to school."

[4] "Village Cadre (*cun liangwei*)" in China refers to a category of officials who play a crucial role in local governance within rural communities. These individuals are typically tasked with implementing government policies, coordinating local development efforts, and maintaining social order. They serve as the link between the central government and rural populations, facilitating communication and the execution of national and local initiatives.

> "Many young people in the village would move to the central arterial road, but do not have enough money."
>
> "Our family has just renovated our mansion. Now the house is quite nice, but there are no shops, entertaining, or health centre."

–Representatives M2801, M4702, F2404 & F3905.

Theme 2: Refurnishing Old Houses

Senior farmers in their 60s or 70s are still attached to their native land. They show respect to their ancestors by preserving and restoring the old houses.

> "People are getting old, and they haven't lived in a big house all their lives. They haven't built the old mansions beautifully to add lustre to ancestors."
>
> "My classmates live in nice, clean and beautiful houses in the city, but dissatisfied, feeling they've left ancestral roots. I, too, am particularly torn."

–Representatives M7303 & F6706.

Theme 3: Reducing Negative Impacts on Health

All have expressed their knowledge about the harm landfills do to villagers' health. Some of them joined protests until things changed in 2019.

> "That landfill is always a strange smell. We dare not open the window. My son says it emits a gas called dioxin. I suspect I have diseases!"
>
> "The odour must be bad for the elderly and the small children! Before 2019, I've been to more than one protest, including online ones."
>
> --Representatives M7303, F6706, M2801, M4702, F2404, and F3905

Theme 4: Co-production of Relocation with Mutual Trusts

Female villagers were first touched by cadres helping in fieldwork. Villagers' requests were mainly accepted and committed, except for unreasonable ones.

> "Our house is newly built and I should be given 3 villas for each of my 3 sons! But the cadres said they now enjoy urban households' benefits."
>
> –Representative M7303
>
> "We, as an old married couple, saw pity in the staff, and let them in the house. When we farmed, they always helped. We were touched."
>
> "My parents lived in my leaked old house. I couldn't make much money. Now I'm given a 3-storey villa and 1.8 million CNY. What a good policy!"
>
> "When you design the villa, please reserve a place for the tractor. As land growers in the village, we need to put machinery."
>
> "Schools for children need to be built. … They actually built an elementary school. It only takes 10 minutes to walk."
>
> "Please design an activity room for the elderly. Now we have a nice big senior room for poker and mahjong. We can even dance in it!"
>
> –Representatives F2404, F3905, F6706, M2801, and M4702

Theme 5: Jobs for Women Close to Families

Later, villagers found that the incineration sites mainly needed male workers as garbage collectors. Women expressed their willingness to have jobs.

> "We (Women) also want to earn money, and don't want to be dependent on men all the time. Why can't we go into the factory?"
>
> "There are so many women in our village, they could open a kiosk for factory workers and tourists to shop near home. Wouldn't that be nice?"

–Representatives F3905, F2404, and M2801.

4.2 Neighbourhood Mechanism Calibrating Synergy

Based on the above deliberation, a circular economy industrial park, where the MSW incineration plant is exhibited with the function of a public environmental education museum, together with a pilot area for rural homestead (*zhaijidi*) reform of Beijing municipality, was established. It relocates seven surrounding villages into two new villa communities. The western zone, or 'West Garden (*xiyuan*) of Yulin', include four villages of Zhanshang, Gaodian, Shantaizi and Zhengfuzhuang. The other three villages of Qianyechang, Houyechang and Tongzhoumafang are in the eastern zone, or 'East Garden (*dongyuan*) of Yulin', close to the largest ancient royal mulberry garden in northern China.

As is integrated into Daxing's urban planning, a commercial zone of Beijing Daxing International Airport is also combined into this 'pilot-park-port' complex with a new variant of 'niche-nexus-nodes' synergy (see Table 3). Thus, the former NIMBY polylemma is enthusiastically turned into YIMBY of shared neighbourhood benefits (*bilin xiaoying*).

5 Discussion and Conclusion

Starting from CDM in China, inspired by the institutional incentive of global climate governance, stakeholders at Anding Town have made good use of technologies from landfill to incineration, with political and capital resources, strengthening and calibrating synergy into an optimal one.

Niches of MSW incineration technologies and homestead policy innovations are embedded into the 'pilot-park-port' complex, achieving visions at various levels, from the Jing-Jin-Ji region to companies at national and village levels. Instead of following traditional demolition resettlement practices, there is no 'land acquisition (*zhengdi*)' or 'transfer to non-farmer identity (*nongzhuanfei*)' in the relocation project. Several top-ranking niches are made as milestones.

Nexus of 'good neighbours' are achieved in the relocation projects by meeting communal voices through face-to-face interactions. Grassroot cadres' hard work and

Table 3 A New "Niche-Nexus-Nodes" synergy in Anding WTE project

Synergy created	Actors	Levels	Type	Provision	Needs
Niche	Leading group for *Jing-Jin-Ji* development	Trans-provincial	Public	Eco-friendly Free Trade Zone Pilot	Regional Integration
	Beijing Dev. & reform committee (BDRC)	Provincial	Public	Rural homestead & collective land reform pilot: 420,000 m^2 of self-occupied or collective houses marketed	Renewable energy rate increased
	Beijing environment engineering group	SOE	Hybrid	Investing 7.03 b CNY for the Circular Park; Benefiting the airport economic zone and Beijing Eco. &Tech. Dev. Zone, or *Yi-zhuang*	Relocation subsidies; Incineration project
	Beijing Urban construction group	SOE	Hybrid	A Museum for Eco-Education; Power (659 m kWh/year) and heat for public use, incl. Daxing Airport	Power grid connection
Nexus	Beijing anxing huimin real estate	SOE	Hybrid	Villas and flats of 101.97 ha, for 2048 hh, are available for commercial usage	A high signing ratio (98.3%)
	Villagers	Individuals	Private	Homestead land; Willingness of relocation	Improvement of public facilities, health, mutual trust, jobs and sense of belonging
	Anding town government	Town	Public	Anding Town Model of relocation: "*Li-fang* Partnership" Mechanism; "Good Neighbour" Council; diverse community activities; Financial compensation, shares and dividends	Beijing's largest MSW incineration site
Nodes	Beijing New Aerotropolis	SOE	Hybrid	Jobs for villagers in neighbourhoods, business chances and for tourists	Heating services for 8 million m^2. Prosperity of the airport economic zone for the *Jing-Jin-Ji* region

Source Authors based on field research, interviews and public news trust-building communications, 'Lifang Mechanism' and rich public services enabled a '*Lifang*' system, having *fang* (20 households) and *li* (10 *fangs*) as administrative units, like in the ancient Han Dynasty of China. An 'Anding Town Model' of relocation, composed of 'government subsidies + rural homestead reform + collective property marketised', is created. Drawing from the household contract responsibility system innovation from China's rural development experience, though the ownership (*suoyou quan*) of rural homestead cannot be altered, their usage (*shiyong quan*) and exchange rights (*jiaoyi quan*) could be separated and empowered to households. The relocated houses could even be commercialised for tourist uses, bringing new income.

Nodes of the 'global–local' network are established through functional connectivity between Anding Town and the southern areas of Beijing, including Yizhuang Economic Development Zone and Daxing International Airport. With the strategic contract between Beijing New Aerotropolis and the Anding circular economy park, the waste-heat long-distance transmission can carry a maximum of 420MW of heat load for the Airport Economic Zone (*Lixian* area in Daxing District). It can provide heating for over 10 million m^2 with 100% green energy, reducing carbon emissions by over 340,000 tons annually (Beijing New Aerotropolis 2023). The green transition idea could thus be broadened to the *Jing-Jin-Ji* region and world travellers.

However, the life cycle social cost of MSW in China is still very high, mainly funded by national subsidies (Jiang et al. 2024; RUC 2017). With the Circular Park open to the public in August 2024, the experience from NIMBY to YIMBY is to be shared in human wisdom-originating measures. The recommended user payment of 1 CNY/day (Jiang et al. 2024; Wang 2020; Chu et al. 2019) might be realised, changing currently impossible into positive outcomes in the near future.

Declaration of Competing Interests This research is supported by the National Social Science Foundation of China (No.: 23GJB00438).

There are no competing interests to report.

References

Arena U (2012) Process and technological aspects of municipal solid waste gasification: a review. Waste Manag 32(4):625–639

Assamoi B, Lawryshyn Y (2012) The environmental comparison of landfilling vs. incineration of MSW accounting for waste diversion. Waste Manag 32(5):1019–1030

Beijing New Aerotropolis (2023) Teaming up with Anding Jinghuan Company, for the largest WTE waste-heat long-distance transmission and heating project in the Airport Economic Zone. China Environment and Sanitation Science and Technology Net. 22 August 2023. Available at: https://www.cn-hw.net/article/detail/900787053896138752. Access on 11 Jan 2025

Bosmans A, Vanderreydt I, Geysen D, Helsen L (2013) The crucial role of waste-to-energy technologies in enhanced landfill mining: a technology review. J Clean Prod 55:10–23

Buchanan JM, Tullock G (1960) The calculus of consent: logical foundations of constitutional democracy. University of Michigan Press, Ann Arbor

Chandra G (2010) Participatory rural appraisal. In: Katiha PK, Vaas KK, Sharma AP, Bhaumik U., Chandra Ganesh (eds) Issues and tools for social science research in inland fisheries. Central Inland Fisheries Research Institute, Barrackpore, Kolkata, India. Bulletin, 163:286–302

Che Y, Yang K, Jin Y, Zhang W, Shang Z, Tai J (2013) Residents' concerns and attitudes toward a municipal solid waste landfill: integrating a questionnaire survey and GIS techniques. Environ Monit Assess 185:10001–10013

Cheng H, Hu Y (2010) Municipal solid waste (MSW) as a renewable source of energy: current and future practices in China. Biores Technol 101(11):3816–3824

Chu Z, Wang W, An Z (2019) Charging for municipal solid waste disposal in Beijing. Waste Manag 94:85–94

Chung W, Choi J, Woo CH, Lee S, Saindon CE (2016) Community relations dealing with a not in my back yard (Nimby) context. IJCMA 3(27):424–452. https://doi.org/10.1108/ijcma-09-2014-0069

Collaborator (2021) Liulitun waste incineration power plant, Haidian District, Beijing, China (Conflict ID: 2412) 14 Oct. https://ejatlas.org/conflict/protest-against-the-plnnaed-liulitun-waste-incineration-power-plant-haidian-district-beijing-china. Access on 19 Jan 2025

Cui J, Liu X, Sun Y, Yu H (2020) Can CDM projects trigger host countries' innovation in renewable energy? Evidence of firm-level dataset from China. Energy Policy 139:1–10. https://doi.org/10.1016/j.enpol.2020.111349

D'Alisa G, Burgalassi D, Healy H, Walter M (2010) Conflict in campania: waste emergency or crisis of democracy. Ecol Econ 70(2):239–249

Das S, Lee SH, Kumar P, Kim KH, Lee SS, Bhattacharya SS (2019) Solid waste management: scope and the challenge of sustainability. J Clean Prod 228:658–678

Dear M (1992) Understanding and overcoming the NIMBY syndrome. J Am Plann Assoc 3:288–300

Demaria F (2021) Asuwei waste incinerator in Changping, Beijing, China. 14 Oct. EJAtlas (Conflict ID: 2825). https://ejatlas.org/conflict/asuwei-waste-incinerator-in-changping-beijing. Access on 25 Oct 2024

Deng Y (2019) The situational legitimacy of social organizations from the perspective of social construction. Social Sciences in China (Chinese Edition) 282(6)

Devas N, Grant U (2003) Local government decision-making–citizen participation and local accountability: some evidence from Kenya and Uganda. Public Adm Dev: Inter-Natl J Manag Res Pract 23(4):307–316

Devine-Wright P (2009) Rethinking NIMBYism: the role of place attachment and place identity in explaining place-protective action. J Commun Appl Soc Psychol 19(6):426–441

Devine-Wright P (2011) Public engagement with large-scale renewable energy technologies: breaking the cycle of NIMBYism. Wiley Interdiscip Rev: Clim Chang 2(1):19–26

Dines N (2013) Bad news from an aberrant city: a critical analysis of the British press's portrayal of organised crime and the refuse crisis in Naples. Modern Italy 18(4):409–422

Dunsire A (1990) Holistic governance. Public Policy Adm 5(1):4–19

EJAtlas, Global atlas of environmental justice. https://ejatlas.org/2025. Access on 19 Jan 2025

Environment Protection Agency, the U.S. (EPA) (2024) Understanding global warming potentials. Greenhouse gas emissions

Fielder HMP, Dolk H, Poon-King CM, Palmer SR, Moss N, Coleman G (2000) Assessment of impact on health of residents living near the Nant-y-Gwyddon landfill site: retrospective analysis Commentary: impact on health needs assessing from different angles. BMJ 320(7226):19–23.4

Fischer F (2000) Citizens, experts, and the environment: the politics of local knowledge. Duke University Press

Fischer F (2012) Participatory governance: from theory to practice. In: David L-F (ed) The oxford handbook of governance. Oxford University Press, Oxford. https://doi.org/10.1093/oxfordhb/9780199560530.013.0032

Gao F (2016) (in Chinese). Revealing Beijing's largest waste treatment plant–Lujiashan waste incineration plant. World Environ (Shijie Huanjing) 4:80–81

Goodnow FJ (1900) Politics and administration: a study in government

Le Grand J (1991) Quasi-markets and social policy. Econ J 101(408):1256–1267
Hare D (1999) Women's economic status in rural China: household contributions to male-female disparities in the wage-labor market. World Dev 27(6):1011–1029
Hayek FA (1944) The road to serfdom. The University of Chicago Press, Chicago
He Y (2006) NIMBY Conflict" and its solution: an analysis based on an urban collective struggle. Publ Manag Res 4:93–103
Heath A, Brinbaum Y (2007) Guest editorial: explaining ethnic inequalities in educational attainment. Ethnicities 7(3):291–304
Henry RK, Yongsheng Z, Jun D (2006) Municipal solid waste management challenges in developing countries–Kenyan case study. Waste Manag 26(1):92–100
Hickel J, Kallis G, Jackson T, O'Neill DW, Schor JB, Steinberger JK, Victor PA, Ürge-Vorsatz D (2022) Degrowth can work–here's how science can help: wealthy countries can create prosperity while using less materials and energy if they abandon economic growth as an objective. Nature 612:400–403
Hobbes T (1651) Leviathan, or the matter, Forme, & power of a common-wealth Ecclesiasticall and Civill, London, Printed for A. Crooke, the Green Dragon in St. Pauls Churchyard
Huang X, Yang DL (2020) NIMBYism, waste incineration, and environmental governance in China. China Inf 34(3):342–360
International Energy Agency (IEA) (2021) Methane and Climate Change. Methane Tracker. https://www.iea.org/reports/methane-tracker-2021/methane-and-climate-change. Access on 20 Oct 2024
Jaffe K (2010) Quantifying social synergy in insect and human societies. Behav Ecol Sociobiol 64:1721–1724
Jiang Y, Leng B, Xi J (2024) Assessing the social cost of municipal solid waste management in Beijing: a systematic life cycle analysis. Waste Manag 173:62–74
Jiang T (2023) Governmental trust, new media use and nimby facility's acceptance: evidence from China. In: 2023 9th International symposium on social science. https://doi.org/10.12783/dtssehs/isss2023/36054.
Johnson T (2013) The Politics of waste incineration in Beijing: the limits of a top-down approach? J Environ Plan Policy Manag 15(1):109–128
Joseph K (2006) Stakeholder participation for sustainable waste management. Habitat Int 30(4):863–871
Kasperson RE, Golding D, Tuler S (1992) Social distrust as a factor in siting hazardous facilities and communicating risks. J Soc Issues 48(4):161–187
Lang G, Xu Y (2013) Anti-incinerator campaigns and the evolution of protest politics in China. Environ Politics 22(5):832–848
Li L (2020) "Green" effects of hybrid actors through carbon trading: cases in Beijing. Glob Trans Proc 1:13–22
Li Y, Koppenjan J, Verweij S (2016) Governing environmental conflicts in China: under what conditions do local governments compromise? Public Admin 94(3):806–822
Li H, Wang J, Yang X, Wang Y, Wu T (2018) A holistic overview of the progress of China's low-carbon city pilots. Sustain Cities Soc 42:289–300
Li W, Ding Y (2016) Report on training in Japan on intensive, economical, and ecological urban landscaping and urban vertical greening construction. Xiamen Botanical Garden. 14 April. Available at: https://xiamenbg.com/Home/newsDetail?typeId=0&newsId=322. Access on 11 Jan 2025
Li X, Ding Y (2020) Holistic governance for sustainable public services: reshaping government–enterprise relationships in China's digital government context, pp 1–20
Lidskog R (1997) From conflict to communication? Public participation and critical communication as a solution to siting conflicts in planning for hazardous waste. Plan Pract Res 12(3):239–249
Lipsky M (1980) Street-level bureaucracy: Dilemmas of the individual in public services. Russell Sage Foundation, New York

Lober DJ, Green DP (1994) NIMBY or NIABY: a logit model of opposition to solid-waste-disposal facility siting. J Environ Manag 40(1):33–50

Lu H (2023) Learning from failure: breaking the waste incineration NIMBY cycle through participatory governance. Clean Waste Syst 5:1–12

Marshall RE, Farahbakhsh K (2013) Systems approaches to integrated solid waste management in developing countries. Waste Manag 33(4):988–1003

Mattiello A, Chiodini P, Bianco E, Forgione N, Flammia I, Gallo C, Panico S (2013) Health effects associated with the disposal of solid waste in landfills and incinerators in populations living in surrounding areas: a systematic review. Int J Public Health 58:725–735

McKay G (2002) Dioxin characterisation, formation and minimisation during municipal solid waste (MSW) incineration. Chem Eng J 86(3):343–368

Nathanson J (2015) Solid-waste management. Available online: http://www.britannica.com/EBchecked/topic/553362/solid-waste-management. Access on 29 July 2023

Nelson N, Wright S (1995) Power and participatory development: theory and practice. ITDG Publishing

Nemerow NL, Agardy FJ, Sullivan PJ, Salvato JA (2009) Environmental engineering: environmental health and safety for municipal infrastructure, land use and planning, and industry. John Wiley & Sons

Nikkhah HA, Redzuan M (2009) Participation as a medium of empowerment in community development. Eur J Soc Sci 11(1):170–176

Njoku PO, Edokpayi JN, Odiyo JO (2019) Health and environmental risks of residents living close to a landfill: a case study of Thohoyandou Landfill, Limpopo Province, South Africa. Int J Environ Res Public Health 16(12):2125

O'Hare M (1977) Not on my block you don't: facility siting and the strategic importance of compensation. Public Policy 24(4):407–458

Ozbay G, Jones MH, Gadde M, Isah S, Attarwala T (2021) Design and operation of effective landfills with minimal effects on the environment and human health. J Environ Public Health, 1–13. https://doi.org/10.1155/2021/6921607

Porta D, Milani S, Lazzarino A, Perucci C, Forastiere F (2009) Systematic review of epidemiological studies on health effects associated with management of solid waste. Environ Health 1(8). https://doi.org/10.1186/1476-069x-8-60

Quantifying social synergy in insect and human societies

Rootes C, Leonard L (eds) (2010) Environmental Movements and Waste Infrastructure London: Routledge

RUC (2017) Evaluation report on the social cost of Beijing municipal solid waste incineration. National Academy of Development and Strategy. Available at: http://nads.ruc.edu.cn/upfile/file/20170322152504_900619_56051.pdf. Access on 10 Jan 2025

Rummery K (2006) Partnerships and collaborative governance in welfare: the citizenship challenge. Soc Policy Soc 5:293–303

Salah MM, Al-Sari' MI, Al-Khatib IA, Kontogianni S (2020) Local residents' perception of landfill impacts in Palestine: the case of Zahrat Al-Finjan landfill. J Mater Cycles Waste Manag 22:673–681

Schiopu AM, Gavrilescu M (2010) Options for the treatment and management of municipal landfill leachate: common and specific issues. CLEAN–Soil, Air, Water, 38(12):1101–1110

Schuftan C (1996) The community development dilemma: what is really empowering? Commun Dev J 31(3):260–264

Scupola A, Zanfei A (2016) Governance and innovation in public sector services: the case of the digital library. Gov Inf Q 33:237–249

Sharholy M, Ahmad K, Mahmood G, Trivedi RC (2008) Municipal solid waste management in Indian cities–a review. Waste Manag 28(2):459–467

Stoica A (2022) A costly joke: global waste recycling. energy industry review. 29 Mar 2022. Available at: https://energyindustryreview.com/environment/a-costly-joke-global-waste-recycling/. Access on 20 Jan 2025

Strauss JC (2009) Forestry reform and the transformation of state capacity in Fin-de-Siecle China. J Asian Stud 68(4):1163–1188
Stringer LC, Reed MS (2007) Land degradation assessment in southern Africa: integrating local and scientific knowledge bases. Land Degrad Dev 18(1):99–116
Temper L, del Bene D, Martinez-Alier J (2015) Mapping the frontiers and front lines of global environmental justice: the EJAtlas. J Political Ecol 22(1):255–278. https://doi.org/10.2458/v22i1.21108
The State Council (2012) The 12th five-year plan on facility construction of urban waste harmless disposal. Beijing, China
Tian C, Han C (2022) How can china resolve the nimby dilemma in a network society? Government and society-negotiated decisions based on evolutionary game analysis. Sustainability 3(14):1308. https://doi.org/10.3390/su14031308
Tomé J (2023) Metropolitan governance in the Argentinian context: The case of the CEAMSE, Doctoral dissertation
Turner SG, Maschi TM (2015) Feminist and empowerment theory and social work practice. J Soc Work Pract 29(2):151–162
UNFCCC (2024) Project 0296: anding landfill gas recovery and utilisation project, 24 Oct 2024, https://cdm.unfccc.int/Projects/DB/DNV-CUK1141654012.19/view. Access on 24 Oct 2024
Van den Akker J (1999) Principles and methods of development research. In: Design approaches and tools in education and training, pp 1–14
Van der Horst D (2007) NIMBY or not? Exploring the relevance of location and the politics of voiced opinions in renewable energy siting controversies. Energy Policy 35(5):2705–2714
Vittes ME, Pollock PH, Lilie SA (1993) Factors contributing to NIMBY attitudes. Waste Manage 13:125–129
Walsh E, Warland R, Smith D (1997) Don't burn it here: grassroots challenges to trash incinerators. The Pennsylvania State University Press, PA
Wang Z, Lin Q, Yin X (2015) What gives Shenzhen the confidence to "burn" a new path? Shenzhen Special Zone Daily (*shenzhen tequ bao*). 9 Oct 2015. Available at: https://news.sina.cn/2015-10-09/detail-ifxirmqz9615171.d.html. Access on 20 Jan 2025
Wang J (2020) As large as 100 football fields, China's largest landfill is nearly full. Xinhua News Agency. 1 Jan. Available at: https://www.xinhuanet.com/politics/2020-01/01/c_1125410881.htm. Access on 10 Jan 2025
Wilson DC (2007) Development drivers for waste management. Waste Manag Res 25(3):198–207
Xu B, Jin L (2022) Responding to the risk of NIMBY conflicts from the perspective of deliberative democracy: based on the case analysis of the resumption project of Wuxi Xidong waste-to-energy Plant. Local Governance Res 2022(02):29–40+79
Young S (2013) Holistic governance. In: Idowu SO et al (eds) Encyclopedia of corporate social responsibility. Springer-Verlag, Berlin Heidelberg, New York, London, pp 1369–1370
Yu H, Qin H (2005) The experiment of public-private partnerships in China. Shanghai: Shanghai Century Publishing (Group) Co., Ltd. (in Chinese)
Zhang W (2007) History of the first CDM project in China. Economy (*Jingji*) 7:22–25 (in Chinese)
Zhang D, Huang G, Xu Y, Gong Q (2015) Waste-to-energy in China: key challenges and opportunities. Energies 8(12):14182–14196
Zhao X, Jiang G, Li A, Yun L (2016) Technology, cost, a performance of waste- to-energy incineration industry in China. Renew Sustain Energy Rev 55:115–130
Zhongyuan News (2024) From "everyone avoids it" to "people queue to visit": The state-owned enterprise behind the transformation of waste-to-energy plants into energy eco-parks. 3 Sep 2024. Available at: https://baijiahao.baidu.com/s?id=1809146070119772430&wfr=spider&for=pc. Accessed on 20 Jan 2025

Harmonising Indigenous and Local Knowledge in Climate Smart Agriculture: Practical Insights from the Satoyama Initiative

Maiko Nishi

1 Introduction

Climate change has been posing global risks onto food supply along with water security and food market operation (Calvin et al. 2023; Lipper et al. 2014). Its greatest adverse impacts, including acute food insecurity, are identified especially in developing and least developed economies and within certain groups of people such as Indigenous Peoples, small farmers and low-income households. Despite the growth of total agricultural productivity aided by technological advancement, climate change has been hindering the progress in productivity improvement over the last five decades globally, but more negatively impacting mid- and low-latitude regions (Calvin et al. 2023). Agriculture is a major economic source in many of these regions and countries, but also one of the major economic sectors most sensitive to climate change (Calvin et al. 2023; Shahzad et al. 2021; Wheeler and von Braun 2013). At the same time, agriculture is a great contributor to global environmental challenges, not only climate change but also biodiversity loss and air and water pollution (Campbell et al. 2017; Yang et al. 2024). It accounts for nearly 12% of global anthropogenic greenhouse emissions alongside great responsibility for excessive water, nitrogen and phosphorus use (Campbell et al. 2017; Chataut et al. 2023; Yang et al. 2024). Given the need to address increasing food demand and malnutrition under the projected global population growth and dietary changes associated with income levels, agriculture will make increasingly negative impacts on the environment unless transformative changes occur in agri-food systems (IPBES 2024; Vishnoi and Goel 2024; Yang et al. 2024).

M. Nishi (✉)
Department of Global Coexistence Studies, College of Bioresource Sciences, Nihon University, Fujisawa, Japan
e-mail: nishi.maiko@nihon-u.ac.jp

M. Okitasari et al. (eds.), *Equitable and Sustainable Cities*, Advances in 21st Century Human Settlements, https://doi.org/10.1007/978-981-92-2793-8_6

In response to these global challenges, the transition toward Climate Smart Agriculture (CSA) has been called for and facilitated by key international organisations such as the World Bank and the Food and Agriculture Organization of the United Nations (FAO) (Akamani 2021; Li et al. 2022; Taylor 2018). As originally defined as 'agriculture that sustainably increases productivity, resilience (adaptation), reduces/removes greenhouse gases (GHGs) (mitigation) and enhances achievement of national food security and development goals' (FAO 2010, p. ii), CSA has been put forward as an integrated approach to strategically managing agricultural systems to address food security, poverty and climate change. In particular, it is intended to achieve 'triple wins' of 'enhanced productivity, resilience and carbon sequestration' (World Bank Group 2012, p. 13).

Since its official proposition by FAO in 2010, the concept of CSA has become increasingly relevant and applied in many international institutions and governments through policies, strategies, partnerships and investments (Akamani 2021; Dinesh et al. 2017; Pinto et al. 2020; Zhao et al. 2023). Following the earlier dialogue and engagement in CSA among key stakeholders, the Global Alliance for Climate Smart Agriculture (GACSA) was established in 2014 as a multi-stakeholder platform on CSA (Alexander 2019; Chandra et al. 2017). Also, FAO has offered a unifying governance framework to diffuse agricultural methods and technologies hinging on 'triple-win' approach, while currently positioning CSA in FAO Strategic Framework 2021–2031 that guides the overall FAO's work (FAO 2021; Taylor 2018). The World Bank has been scaling its engagement and investment in CSA, placing CSA as a core of all the World Bank's new operations in the agriculture and food sectors (The World Bank 2024). A significant potential of CSA is envisaged with data-oriented strategies in combination with recent remarkable technological advancements, which can disrupt conventional farming practices and improve agricultural efficiency, productivity and profitability in a sustainable manner (Thomas et al. 2023). In this context, the CSA market size is projected to reach USD 25.4 billion by 2028 globally with a 9.4 per cent growth rate from 2023 (Thomas et al. 2023).

However, in the absence of a clear definition and criteria in regard to what practices or actions could lead to the 'triple wins', CSA has received considerable criticisms even from the outset of its evolution (Karlsson et al. 2018). Besides uncertainty about the practical realisation and operation of CSA (Pinto et al. 2020; Taylor 2018), major criticisms converge on its contradictions to sustainability, ranging from negative effects on biodiversity, inequitable burdens on smallholder farmers, promotion of genetically modified crops degrading genetic diversity and local cultures, and greenwashing by private interests to further land grabs by corporate oligarchy and political elites (Alexander 2019; Chandra et al. 2017, 2018; Karlsson et al. 2018). As urged by the critiques of CSA, it is imperative to take into consideration local socio-political contexts and agro-ecological diversity, particularly through a bottom-up approach centred around farmers, and to identify technologies and supporting measures that are adaptable and appropriate to the specific contexts (Alexander 2019; Huyer and Partey 2020; Karlsson et al. 2018; Wakweya 2023).

In this vein, the role of Indigenous and local knowledge (ILK) in CSA has become highlighted to enhance resilience to climate change and attain sustainable development (Akamani 2021; Amare and Gacheno 2021). Family farmers and local communities have accumulated abundant knowledge over generations through experiencing harsh climate conditions, experimenting with locally invented methods through trial and error, adapting to challenges including climate-related problems and fostering resilience to such challenges that vary within spatial, temporal and institutional contexts (Amare and Gacheno 2021; Mafongoya and Ajayi 2017). Such knowledge, interchangeably called, for instance, traditional knowledge, traditional ecological knowledge and Indigenous knowledge, has been used in many domains of climate action and sustainable agricultural development, including climate prediction and climate-resilient methods for farming, water use and food security (Amare and Gacheno 2021). The usefulness of ILK in planning and developing strategies for climate adaptation and sustainable development has been increasingly recognised (Mafongoya and Ajayi 2017). In particular, given the complexity of agri-food systems embedded in dynamic human-nature interactions, the beneficial use of diverse knowledge systems where ILK is also harmonised or paired with modern science has been gaining attention in CSA initiatives (Akamani 2021; Alexander 2019). Yet, literature on ILK harmonisation in CSA is scarce, whereas studies on CSA adoption remain limited particularly in regard to its implications for food security of small farmers (Wakweya 2023). Also, such knowledge pairing is likely to face challenges like different worldviews, different methodological approaches, power asymmetries and a lack of trust among stakeholders (Akamani 2021).

This chapter explores how ILK can be harmonised or effectively paired with modern science and technologies for CSA, drawing on the case studies collected through the International Partnership for the Satoyama Initiative (IPSI). Launched in 2010, IPSI has been fostering a global platform where a diverse range of member organisations (e.g. national and local governments, NGOs, academia, private sector, Indigenous Peoples and local communities) are dedicated to working together to achieve the vision of the Satoyama Initiative 'societies in harmony with nature' (IPSI 2019). Under this vision, the Satoyama Initiative is committed to taking the three-fold approach to sustainable management of productive landscapes and seascapes. This approach entails the following three interrelated steps: (1) consolidating wisdom on securing diverse ecosystem services and values; (2) integrating traditional ecological knowledge and modern science to promote innovations; and (3) exploring new forms of co-management systems or evolving frameworks of 'commons' while respecting traditional communal land tenure (IPSI 2020). In such a way, knowledge pairing or coproduction is a key to engaging in the Satoyama Initiative, although knowledge is not specifically rendered for CSA but broadly for managing landscapes and seascapes where production activities (e.g. agriculture, forestry and fisheries) are conducted. As part of this effort, IPSI has been collecting and sharing the case studies submitted by its members on their landscape and/or seascape management.

Based on the review of IPSI case studies, this chapter aims to address the following questions: (1) What ILK is used or applied for CSA, (2) How such knowledge is harmonised or paired with modern science and new technologies and (3) What effects are derived from the use or application of ILK to achieve 'triple wins' of CSA. The next section details the methodology for the study, which is followed by the review findings with respect to each of the above three questions as well as implications for disruptive innovations for sustainability. The chapter ends with a conclusion where practical implications are drawn out.

2 Methodology

This study used the IPSI case study database (https://satoyama-initiative.org/case_study/#start), which is openly available, to identify and extract the IPSI cases where ILK is applied and/or paired with modern science in planning and implementing CSA. Within the database, containing 320 case studies as of 30 November 2024, relevant cases were searched, using the following keywords: (Climate) AND (Technolog* OR Knowledge) AND (Traditional OR Local) AND (Agriculture OR Farm*). This resulted in the identification of 113 case studies, which were submitted between May 2010 and May 2024 to the IPSI Secretariat hosted by the United Nations University Institute for the Advanced Study of Sustainability (UNU-IAS). Out of these cases, the review focused on 82 cases submitted after 2014 to determine more recent status and trends over the past decade. The provisional review was conducted to ensure the relevance to climate action, application or pairing of ILK and agricultural practices, by focusing on the title, summary or abstract and introduction section of each case study. This process led to the extraction of 29 case studies. Then, the full-text review was conducted based on the inclusion criteria including: (1) concrete interventions to address climate change, (2) relevance to agricultural production activities (narrowly defining these activities by excluding, for instance, forest restoration and marine and coastal management) and (3) sufficient information about ILK used or applied. This review resulted in the selection of 2 case studies.

The detailed review of the selected two case studies (Table 1) drew on the comparative lens to identify commonalities and distinctions between the two in terms of the local settings, the knowledge used, the approaches and processes of knowledge pairing and creation, and the effects for CSA. The effects of CSA were analysed with respect to the triple-win goals of CSA (i.e. agricultural productivity, climate resilience and adaptation and climate mitigation). To draw out implications of the two case studies for a broader range of local contexts, insights from the screening of the 29 potentially relevant case studies as well as literature review were also used for the analysis.

3 ILK Application for CSA in IPSI Case Studies

Given the vision 'societies in harmony with nature' in the Satoyama Initiative, IPSI entails a broad range of activities related to the management of productive landscapes and seascapes, including capacity development, research, policy contributions, networking and communications. In this regard, the purposes of ILK use or application include a wide range of land/seascape management activities such as ecosystem restoration, protected area management and ecotourism, not necessarily for CSA practices in a narrow sense (excluding forestry and fisheries). The screening of the 29 potentially relevant case studies reveals that ILK in combination with modern science has been applied or identified as useful for various purposes related to climate action and land/seascape management – including mangrove rehabilitation for climate resilience in the Philippines (Gevaña et al. 2021), woodland restoration involving fire risk management in Ghana (Mbawine and Dzekoto 2023), recovery of chestnut cultivation effective for local climate regulation in Spain (Díaz-Varela et al. 2018), and sustainable rural development in the forest and Milpa landscape serving as a key carbon reservoir in Mexico (Méndez-López et al. 2022). In these cases, while using scientific knowledge or modern technologies (e.g. provision of technical or mechanical tools, new research results), ILK was applied or considered useful for local or regional solutions to address climate challenges and help attain the vision. Such knowledge included certain features and effects of species and genetic resources, traditional management practices and social organisational or governance mechanisms.

Among the two selected cases, one is concerned with the restoration of rice paddies and associated ecosystem services in Wayanad District, Kerala Province, India (Bhatt et al. 2023), and the other is focused on the smart agriculture initiative in Zhongliao community of Tamsui District, New Taipei City, Taiwan (Wang et al. 2023). The former case study in Wayanad features Rice Seed Village (RSV) programme of MS Swaminathan Research Foundation (MSSRF). This initiative started in 2011 as a participatory rice seed production programme to conserve genetic diversity of rice crops at the maximum possible level and build capacity of rice cultivators for climate change adaptation in five villages of Wayanad District. Prior to the launch of this programme, Wayanad had experienced, over several years, various climate variability (e.g. frequent floods and erratic rainfalls during monsoon seasons and drought during summers). At the same time, it was also going through enormous agroecosystem changes associated with land use changes, modern agricultural practices, other anthropogenic activities and climate change. In response, the programme aimed at biodiversity conservation, climate change adaptation and livelihood development.

The latter case study in Zhongliao presents a multi-stakeholder effort to incorporate green energy into smart agriculture and develop a circular socio-economic and ecological model in Zhongliao community. This effort evolved from the community's initiative that established a platform in 2013 in collaboration with different ministries and governmental councils to jointly develop plans for addressing environmental degradation and protecting the residents' health in the community. Affected by urbanisation and industrialisation since the 1960s, the community had experienced rural exodus (particularly the outmigration of the young population), the ageing and shrinking of the farming population and a decline in the labour force. Also, land degradation due to the expanded use of agrochemicals and the establishment of copper manufacturing factories as well as aquaculture and pig farms upstream led to the pollution of rivers and air and the deterioration of ecological environment in the community. These trends slowly led to land abandonment. In this context, the concept of the Satoyama Initiative was introduced to the Agriculture Bureau of New Taipei City, the Soil and Water Conservation Bureau and the community residents, aiding the development of eco-friendly farming and related activities. Meanwhile, government policies were promoted for energy conservation, carbon reduction and green energy development.

In both cases, ILK was combined with modern science to address the challenges and achieve the objectives of the initiatives. In the Wayanad case, the programme began with the documentation of traditional knowledge and practices by mobilising the rice farming families interested in cultivating traditional rice varieties (TRVs). This was followed by the validation of the derived knowledge among multiple knowledge holders during the farmers' training sessions that started in 2015. For instance, the discussions among farmers and agriculture experts revealed that the traditional practices adhered to the lunar cycle (called *Paurnami*) for sowing, transplanting and harvesting are effective for pest control because rodents and other pests avert themselves from the fields for certain periods in the lunar cycle. Also, many other traditional practices appeared to be eco-friendly in managing diseases and pests and reducing dependency on agrochemicals while maintaining the fabric of natural food web. Examples include using the opposite leaf Fig (*Ficus hispida*) for rebuffing insect larvae from the fields due to its rough and spiny leaf surface, attracting owls as a natural predator of pests (e.g. rodents), using stuffed owl feathers to avoid rodent-caused damage in the fields, and using cow dung as fertilisers. The cultivation of TRVs with more height and thicker straw also appeared to be suitable for cattle grazing in the fields after the harvest, rendering the remaining stubble for grazing and promoting *in-situ* manuring. Local governance within the community also aided the equitable distribution of manure by demarcating the grazing fields and shifting cattle across the borders over time.

In the Zhongliao case, traditional wisdom was applied to revitalise the productive landscape with organic and eco-friendly farming, where new agricultural knowledge was gradually introduced by the Agriculture Bureau of New Taipei City. Under the community association's initiative to attract young people to return and engage in organic and eco-friendly farming, the young farmers of Yuanfeng Organic Farm conducted experiments on crops and soil to improve the quality of soil that had been acidified by farming. Building on traditional knowledge and practices, this effort led to the use of the fermented waste produced through husbandry as organic fertiliser for crops and feed for laying hens, creating new eco-friendly farming where the cyclic process of organic planting and husbandry was formed. The community also started in 2016 to engage in applying the concept of Aquaponics (i.e. a food production system coupling aquaculture with hydroponics) based on past ecological restoration experiences. By using the growth characteristics of Gramineae or Poaceae, this initiative developed a new system to purify wastewater introduced from the upstream fish farms to the nearby fields for irrigation, whereas fish excretion was sourced for nitrogen fertiliser for plants. The purified water was brought into the adjacent river, improving the ecological status of other species and the water quality downstream.

Moreover, this cyclic approach to water purification and farming was further applied in response to government policies of energy conservation, carbon reduction and green energy. In collaboration with the Agriculture Bureau of New Taipei City and the fishery industry, the community association developed a solar-powered pumping system to pump up the sludge, which was diverted from the upstream fish ponds for purification into the downstream water bamboo field, and then used the purified bottom sludge as farmland fertilisers. This effort contributed not only to water source protection, ecological conservation and air purification but also to the community's additional income earned by harvesting and selling healthy agricultural produce. The discussion within the community, including the returned youths led to the use of some eco-friendly farming products for agri-food education and elders' meals in the community, attaining community public welfare. Furthermore, the community joined forces with Tamsui Cultural Foundation in 2020 to build the New Taipei City Smart Green Energy Community Cooperation to promote citizen power plants for green energy, whereas the solar panels were installed on the roofs rather than the land to avert ecological impacts and allow for productive activities in the community.

4 Approaches and Processes for Harmonising ILK in CSA

Both cases took the community-centred step-by-step approaches to the initiatives where ILK was applied in combination with modern science and technologies, although different stakeholders took the initiative. In the Wayanad case, as mentioned above, MSSRF (https://www.mssrf.org/), a not-for-profit trust based in Chennai, India, started the initiative in 2011. Having engaged since 1988 in promoting the use of modern science for sustainable agriculture and rural development, MSSRF was concerned with the important roles of the rural and Indigenous community in enhancing ecosystem sustainability, mitigating climate change effects and improving livelihoods. Based on the study led by MSSRF that found abundant TRVs cultivated in the Wayanad district in 2011, it proposed the idea of holistically supporting the sustainable provision of ecosystem services and biodiversity conservation in the paddy agroecosystem in the district. Following the community mobilisation and consultation that identified diverse challenges for traditional cultivation and the need for awareness raising and restoration, the participatory programme began with five villages in the district in 2011 initially, but over a decade, expanded to involve 30 villages.

Beginning with community mobilisation, a series of activities were undertaken for the programme in a participatory and stepwise manner. Among others, three key interventions enabled significant impacts of the programme in achieving its objectives. First, 27 training programmes were implemented since 2015, reaching out to over 700 farmers (over 450 males and 270 females). These programmes provided training on various climate resilient and/or adaptation practices and preparation and use of various organic nutrient formulations. For instance, as climate-resilient practices, the System of Rice Intensification (SRI)[1] and the Modified SRI (MSRI) were introduced, allowing farmers not only to cultivate paddies under water scarcity and adapt to drought conditions but also to attain 20 per cent higher yields than conventional methods. Other practices to mitigate and adapt to climate effects included the cultivation of pulse and legume crops in rice fields after the harvest, green manuring, nursery establishment in uplands and change in sowing and/or transplanting dates under delayed monsoon seasons. Along with the use of organic nutrient formulations (e.g. *Jeevamrutham*, *Beejamrutham* and *Fish Amino*), these practices also enhanced yields without harmful chemicals. Further, as mentioned above, ILK was documented and validated in the training course as part of the overarching programme.

Second, a strategic framework was developed in a participatory and experimental manner for conservation, consumption and marketing of TRVs' grains and seeds. Based on the identification of the maximum genetic variability of TRVs known to farmers, the testing and selection of TRVs for cultivation were jointly conducted by farmers and the MSSRF team on site, in consideration of farmers' preferences such as taste, nutritional and aromatic properties and climate tolerance of the varieties. Given that ensuring an adequate supply of quality seeds prior to linking to the market is crucial to harnessing the benefits and special properties of TRVs, participatory seed exchanges between farmers were undertaken along with the abovementioned training programmes. The training for quality seed production included a selection of pure line panicles, exclusion of off-type seeds and effective seed storage methods with the supply of 88 metal storage boxes to the villages (for disease and pest prevention and post-harvest management). Then, marketing channels were developed through a Farmer Producer Organisation (FPO) named WAMPCo. to link the participating villages to the market and facilitate the sale of grains. Thus, the previous informal system of seed exchange was renewed into a more structured one to ensure conservation and quality production of TRVs.

Third, annual seed festivals (called *Vittutsavam*) began in 2014 with support from local and grassroots organisations alongside the participation of the farming communities. This served as a platform or medium for more extensive seed exchange, knowledge sharing and awareness raising across the state of Kerala. Seed exchange expanded statewide, facilitating the exchange of seeds to which farmers were not easily accessible in the past because of the lost market for sale. Participated by over a thousand visitors with diverse backgrounds and expertise each year, this

[1] SRI has been proposed as an integrated eco-frendly approach to rice cultivation, which was developed from participatory on-farm research in Madagascar but has been extended worldwide due to its benefits for CSA (Dobermann 2004; Thakur and Uphoff 2017).

annual festival, consisting of a variety of events such as seminars, panel discussions and exhibitions, allowed for knowledge exchange and awareness raising on diverse topics—including agro-biodiversity, needs for community seed banks, resilient agro-food systems, farmers' rights, government policies and governance and partnerships. In particular, farmers' networking with the association of traditional seed growers allowed for a better understanding of their rights and increased peer recognition, enhancing their strategic decision-making on paddy cultivation, seed conservation and sale and marketing of their products. Furthermore, two cash awards were given to the Indigenous community at each festival to recognise their efforts in agro-biodiversity conservation, motivating them to further engage in TRV cultivation in a biodiversity-friendly manner.

In the Zhongliao case, the initiative emerged in 2013 from the community concerned about local socio-economic and environmental challenges but increasingly involved a broader range of stakeholders over time. In the beginning, the community formed a task force of seven members to hold weekly meetings to discuss community issues, which quickly expanded to involve more than 30 participants in a year. At the same time, the community started to participate in the Rural Rejuvenation Project based on which a variety of stakeholders came to join forces in planning and implementing new activities. Among others, three key areas of intervention can be highlighted for the remarkable impacts of the initiative. First, circular agriculture was promoted as a strategy to recall young people through their engagement in organic and eco-friendly farming in the community. Under this initiative, as earlier mentioned, the young farmers of the Yuanfeng Organic Farm pioneered experimenting and developing new eco-friendly farming where traditional knowledge and practices were applied along with the new agricultural knowledge. This was followed by a variety of eco-friendly farming practices and associated activities, including the application of Aquaponics for water purification and farming, chicken manure composting, agricultural and leisure tourism and conservation education.

As part of this initiative, the resilience assessment workshop was held in 2021, drawing on the Indicators of Resilience in Socio-ecological Production Landscapes and Seascapes (SEPLS) (hereafter SEPLS Resilience Indicators) developed throughout the IPSI efforts (UNU-IAS et al. 2014). This workshop with the involvement of multiple stakeholders (e.g. Agricultural Bureau, enterprises and other relevant units) furthered the comprehensive approach to the development and implementation of a circular agricultural model by combining different types of resources through multi-stakeholder engagement (including industries, government agencies, schools) and integrating diverse educational concepts into ecological conservation. As a result of the effort over the years, 6 young farmers returned to the community during the period between 2013 and 2021 to engage in organic and eco-friendly farming, whereas 16 wetlands based on the Aquaponics concept were developed to restore ecological resources. During this period, the area of eco-friendly farming also increased from 2 to 13 ha, along with the restoration of some key species (e.g. rhinoceros beetles, stag beetles, *Pyrops watanabei*). Furthermore, community environmental education came to involve many more enterprises and schools.

Second, rural–urban reconnection was promoted by developing the farmland renting model to turn abandoned fields into technological farms. Based on the concept of circular rural–urban interlinkages through ecosystem services, the community encouraged the elderly to outsource farming and rent out farmland to urban residents who could engage in eco-friendly farming. This initiative resulted in the revitalisation of abandoned farmland where 180 urban residents participated throughout their holidays in experiencing agrarian life. The earnings from this intervention (more than NTD 500,000/ approx. USD 15,400 each year) were returned to the community, whereas some of the harvested vegetables were given to the North Coast Social Welfare Center as a donation for public welfare. More recently, new technologies were adopted to this model where urban residents can digitise and visualise the field environments and remotely grasp climatic and crop growth images so as to effectively monitor and manage farmland.

Third, green energy was promoted, as earlier mentioned, in response to government policies. Under the leadership of the Chairperson, the community engaged in green energy development and established its first citizen power plant in July 2020. With support from the landowners who offered their land free of charge for investment, 7 power plants came to be established by 2022, generating a total of 416,911kWh that was sold to Taipower (i.e. an electric power industry). Out of the revenue from this sale of electricity, 5% was earmarked for the community funds for elderly care. In combination with all the above-mentioned efforts (e.g. water purification, organic and eco-friendly farming), this intervention also aided in creating an ecologically sound environment in the community, as evident from the ecological survey that identified a total of 842 animals and insects (birds, amphibians, reptiles, fish, shrimps, crabs and butterflies) across 70 species in a sampling area of the community—including 3 rare protected wild species, 9 endemic species of Taiwan and 9 endemic subspecies of Taiwan.

5 Effects for 'Triple Wins' of CSA

5.1 Agricultural Productivity

For both cases, a straightforward analysis of agricultural productivity, such as total factor productivity (TFP) (i.e. a measurement for the amount of agricultural output derived from the combined set of land, labour, capital and material resources used in farm production), is not available. Related to agricultural production, however, the initiatives in both cases appear to be effective in terms of: (1) increased farmers' engagement in ecologically friendly farming, (2) added values of agricultural produce and (3) creation of alternative livelihood means. First, the two cases show that an increasing number of farmers came to be involved in the introduced agricultural practices; accordingly, the areas for such initiatives expanded. In the Wayanad case, 340 farm families came to engage in the cultivation of TRVs along with the increase of the beneficiary villages from 5 in 2011 to 30 in 2021, while the programme reached out to more than 3000 farmers through various activities. In the Zhongliao case, the progress in organic and eco-friendly farming over the decade included more farmers engaging in such practices (e.g. 6 returned young farmers engaging in eco-friendly agriculture, 10 farm households composting chicken manure, 2 organic certified farms, 3 agritourism farms, 5 farms with green conservation label) and an increase in eco-friendly farming area from 2 to 13 ha (including the revitalised abandoned land).

Second, additional values of agricultural produce are also observed through the initiatives in both cases. In the Wayanad case, the intervention allowed for better prices of certain quality seeds and gains (i.e. Veliyan, Adukkan, Thondi, Chennelu, Chomala, Kalladiaran and Chenthadi varieties) with INR 34 (approx. USD 0.42) per kg from the Wayanad Rice Mill and some interested Farmer Producer Organisations (FPOs), compared to INR 24 (approx. USD 0.30) per kg from the open market and smaller mills. In particular, three aromatic rice varieties (i.e. Gandhakashala, Mullankaima and Jeerakashala) gained premium prices of INR 70–110 (approx. USD 0.86–1.35) per kg, roughly 40 per cent more expensive than other varieties, along with the enhanced market linkages and the increased demands associated with the raised consumer awareness. In the Zhongliao case, the initiative brought about additional income for the community by selling the agricultural products (e.g. vegetables, water bamboo) harvested from the newly developed circular agriculture, while new processed products (e.g. pumpkin curry, pumpkin egg rolls, sweet potato Swiss roll) were developed in collaboration with universities through the project.

Third, alternative livelihood means were developed in both cases. In the Wayanad case, the alternative uses of paddies for the cultivation of fish, vegetables and pulses were promoted during the fallow seasons, providing additional livelihood opportunities and food security. In addition, certain paddy fields were used during the summer months as football playgrounds, creating recreational opportunities for youths. In the Zhongliao case, besides green energy development and organic and eco-friendly farming that provided additional revenues for the community, environmental education and ecological tours were promoted, taking advantage of environmental and cultural landscape resources in and around the community (e.g. rich biodiversity, surrounding mountain scenery, smart tea gardening, terraced fields, stone masonry). Among others, various educational and cultural activities (e.g. children's tours on the complete agro-food processes, diverse educational courses for students and insect-watching activities) were conducted in collaboration with surrounding schools. Overall, the participants in community environmental education reached more than 2,000 people annually through the creation of 16,000 leisure tour visits per year, while local employment increased in the community from 20 in 2013 to 43 in 2021 (including the returned youths).

5.2 Climate Resilience and Adaptation

As specifically intended as one of the key interventions, the Wayanad case study clearly shows the effects of climate resilience and adaptation made through the initiative. However, the Zhongliao case entails few focused interventions for climate resilience and adaptation per se while improving farmland microclimate environments that could be adaptable and resilient to climate change in future. With the smart remote climate monitoring technology recently adopted through the effort of rural–urban reconnection, climate adaptability and/or resilience might have increased. Nevertheless, particular effects in this regard are not highlighted in the case study.

In the Wayanad case, various climate resilient and/or adaptation practices were introduced through the training sessions. In particular, the TRVs tolerant to flood and drought conditions known to farmers were promoted as part of the programme. Under the recurrent floods and other climate variabilities, the adaptive and resilient qualities of these varieties were proved, and thus, minimum price depreciation under climate calamity was ensured due to the sustained grain quality. For instance, the Kalladiaan variety, which was promoted as a promising climate-resilient crop, enabled the farmers to grapple with the aftermath of heavy rainfalls during monsoon seasons while making up the losses in other varieties when cultivated after the rainy seasons. Also, seed exchanges helped the communities to recover from the damage in the aftermath of recurrent floods during 2018 and 2019 by complementing the loss of certain varieties damaged. This was demonstrated by the significant increase in volumes of seed exchanges for the three years: 170.2 kg in 2018, 1279.35 kg in 2019 (after the floods) and 767.2 kg in 2020. The seeds of these TRVs were also supplied to farmers in other regions concerned with unexpected weather incidents.

5.3 Climate Mitigation

Generally moving toward sustainable agriculture, on the one hand, the Wayanad case may have contributed to climate mitigation through eco-friendly management practices (e.g. disease and pest control without agro-chemicals, green manuring) and restoration of agro-ecosystems (e.g. soil conservation, nurseries establishment in uplands). Yet, without clearly focused interventions for climate mitigation, definite or measurable effects in this regard are not provided within the case study. On the other hand, with a clear focus on climate mitigation, the Zhongliao case demonstrates the generation of low-carbon energy through the development of citizen power plants along with energy-efficient water purification and organic fertiliser production. In particular, based on the conversion of ecosystem service values, the study shows that the development of 7 citizen power plants resulted in a reduction in a total of 264,812.1 kg of CO_2 emissions, equivalent to saving 22,939 trees.

6 Disruptive Technologies for Transformative Change in Agri-Food Systems?

As mentioned earlier, transformative changes in agri-food systems are urgently required to address current global environmental crises and advance sustainability, wherein disruptive technologies may play a significant role (Schuelke-Leech 2021). The report on transformative change by the Intergovernmental Science-Policy Platform on Biodiversity and Ecosystem Services (IPBES) highlights the importance of embedding technologies in transformative frameworks that facilitate fundamental, system-wide shifts in views, structures and practices in accordance with the four key principles, including (1) equity and justice, (2) pluralism and inclusion, (3) respectful and reciprocal human-nature relationships and (4) adaptive learning and action (IPBES 2024). Furthermore, the IPBES report emphasises that '[w]eaving together insights from diverse approaches and knowledge systems, including Indigenous and local knowledge, enhances strategies and actions for transformative change' (IPBES 2024, p. 6). Conversely, it points to 'limited access to clean technologies and uncoordinated knowledge and innovation systems' as one of the five systemic challenges that impede or prevent transformative change while stressing 'disconnection from and domination over nature and people' as one of the three key underlying causes of biodiversity loss and nature's decline (IPBES 2024, p. 5, 6). In this regard, insights on how ILK has been harmonised or effectively coordinated with modern science and technologies in planning and implementing CSA are indeed relevant and valuable in promoting transformative changes in the agriculture and food sectors toward a just and sustainable world.

The two cases presented in this chapter do not show the adoption of state-of-the-art or ultra-advanced technologies, though the Zhongliao case somewhat demonstrates the use of relatively new modern technologies (e.g. smart remote climate monitoring technology) in the initiatives. Rather, both cases explicate how different kinds of knowledge and technologies, including ILK, have been woven together through the inclusive and collaborative multi-stakeholder processes for equitable benefits from the interventions (e.g. farmers' livelihoods, elderly welfare) while enhancing stakeholders' capacities through learning by doing in the efforts to realise societies in harmony with nature. In particular, these cases underscore the significance of partnerships between different stakeholders, through which local initiatives can be scaled out (i.e. extending the coverage of the initiatives), scaled up (i.e. influencing policies and institutions, including markets) and scaled deep (i.e. impacting views and values) (Moore et al. 2015).

For instance, the Wayanad case shows not only the increase in the participating villages and farmers but also the changes in the TRV market along with the increased demand for TRVs that became more valued by users. The Zhongliao case also illustrates the evolving progress in the initiatives in terms of the number of activities and participants, while demonstrating how energy policies can be implemented on the ground under new institutional arrangements for environmentally friendly energy generation and sale. Although the Zhongliao case does not directly indicate the shift in views and values through the initiatives, the increase in returned youths and their engagements in eco-friendly farming imply their shifts in preferences and values on human-nature relationships. Importantly, given that cities, highly dependent on ecosystem services (e.g. food, water, energy) from other areas, cannot alone create a just and sustainable world, the initiatives for rural–urban reconnection may be impactful for shifting the views and values held by urban residents, many of whom tend to be disconnected from nature. In this respect, new technologies, such as remote climate monitoring technology, could help in provoking spiritual human-nature reconnection and integrating rural–urban interlinkages into views, structures and practices across scales.

7 Conclusion

Given the broader vision 'societies in harmony with nature' in the Satoyama Initiative, only a few of the IPSI case studies exemplify CSA specifically. However, the identified two case studies provide useful lessons and implications for how to effectively apply or harmonise ILK in promoting CSA. They demonstrate that some ILK (e.g. farmers' ecological knowledge related to TRVs, local community's experiential knowledge of ecosystem restoration) aided in promoting CSA and making effects on certain aspects of its triple-win goals (e.g. tolerance to floods and droughts, energy efficient agriculture). Such ILK appears to be effective for CSA if it is appropriately paired with modern science and technologies. Albeit the origins of those initiatives differ in their specific local contexts, the two cases share several common approaches to pairing ILK with modern science and technologies to achieve their goals. First of all, the review of these cases suggests that community mobilisation and consultation should be a key first step in developing interventions. This is essential to identify local problems, gaps in knowledge and existing capacities within a community, build consensus at the local level and develop locally suitable solutions.

Furthermore, given the uncertainties and unpredictability associated with generating new knowledge and technologies for CSA, the process of experimenting, testing and learning is also needed. In this process, both cases show that participatory and multi-stakeholder approaches are practically advantageous to mobilise different types of resources (e.g. technologies, skills, expertise, labour, land, finance), share and disseminate information and knowledge and develop synergistic solutions for CSA. In this regard, on the one hand, an open-ended format of events (e.g. seed festivals, eco-tourism activities) helps with broader dissemination and collaboration in promoting the initiatives. On the other, a strategic or focused format of consultation or assessments (e.g. farmers inventory, ILK documentation, task force weekly meetings, resilience assessments) helps identify particular needs and preferences and develop a niche strategy for specific local solutions. Related to this, the SEPLS Resilience Indicators could be a useful tool to mobilise local communities, assess the status and trends of resilience in land/seascapes holistically and devise integrated solutions, as they are framed in a comprehensive manner to allow for participatory community-based resilience assessments. In addition to the Zhongliao case, many of the 29 cases, which were identified in the case screening process, demonstrate the use of the SEPLS Resilience Indicators which ILK was drawn on for pursuing their initiatives such as ecosystem restoration, eco-tourism and other types of land/seascape management (e.g.Méndez-López et al. 2022; UNDP 2017).

Beyond the specific triple-win goals of CSA, the two case studies present multiple benefits from managing productive landscapes, including biodiversity conservation, water purification, public welfare and educational, cultural and recreational activities. However, the effects of triple-win targets are not necessarily clear or measurable unless specific targets relevant to these aspects are set out for the initiatives. There may be positive or even negative consequences of the interventions in regard to the effects for CSA, wherein trade-offs and synergies may exist or emerge. To ensure the beneficial application and pairing of ILK with modern science and technologies for CSA, clear triple-win targets should be established. To do so, a systematic monitoring and evaluation framework would need to be developed, whereas novel technologies could be adaptably used in accordance with the local contexts for broadly promoting transformative change in agri-food systems.

8 Declaration of Competing Interests

The author has no conflicts of interest to declare that are relevant to the content of this chapter.

Table 1 Selected two case studies

Case study title	Restoring Rice Paddies and Rice Agro-Ecosystem Services through a Participatory Seed Conservation and Exchange Programme	Incorporating Green Energy into Smart Agriculture to Create Regional Circular and Ecological Spheres (R-CES)
Authors	Archana Bhatt, N. Anil Kumar, C. S. Dhanya, P. Vipindas	Chao-Wen Wang, Kuang-Chung Lee, Hsin-Hsun Huang, Ding-Wun Chen
Case study submission	2023	2022
Study sites	• Location: Wayanad District, Kerala Province, India (11.6854 N, 76.1320 E) • Geographical area (ha): 213,100 • Population (persons): 51,320	• Location: Zhongliao community of Tamsui District, New Taipei City, Taiwan (25.1976 N, 121.4775 E) • Geographical area (ha): 313 • Population (persons): 1,300
Keywords	Rice Seed Village, Paddy agro-ecosystem, Participatory seed improvement, Seed exchange, Biodiversity conservation	Socio-ecological production landscapes and seascapes (SEPLS), Biodiversity, Eco-farming, Green products, Participatory Guarantee System, Sustainable rural development
Programme/ Project	Participatory rice seed production programme (since 2011)	Rural Rejuvenation Project (since 2013)
Objectives	• Biodiversity conservation: Conserve the maximum possible rice crop genetic diversity • Climate change adaptation: Build capacity amongst rice cultivators to adapt to climate change and related vagaries • Livelihood development	• Conservation and sustainable use of landscape resources for harmonious human-nature coexistence: - Establish community partnerships (cultivating local talents and developing cross-sector cooperation - Recall young people through the terraced wetland utilisation and circular agriculture - Transform abandoned fields into technological farms for rural–urban reconnection - Develop smart green energy communities

References

Akamani K (2021) An ecosystem-based approach to climate-smart agriculture with some considerations for social equity. Agronomy 11(8). https://doi.org/10.3390/agronomy11081564

Alexander S (2019) What climate-smart agriculture means to members of the Global Alliance for climate-smart agriculture. Future Food: J Food, Agric Soc 7(1):21–30. http://hdl.handle.net/123456789/11264

Amare G, Gacheno D (2021) Indigenous knowledge for climate smart agriculture—a review. Int J Food Sci Agric 5(2):332–338. https://doi.org/10.26855/ijfsa.2021.06.019

Bhatt A, Kumar NA, Dhanya CS, Vipindas P (2023) Restoring rice paddies and rice agro-ecosystem services through a participatory seed conservation and exchange programme. In: Nishi M, Subramanian SM (eds) Ecosystem restoration through managing socio-ecological production landscapes and seascapes (SEPLS). Springer Nature, pp 137–155. https://doi.org/10.1007/978-981-99-1292-6_8

Calvin K, Dasgupta D, Krinner G, Mukherji A, Thorne PW, Trisos C, Romero J, Aldunce P, Barrett K, Blanco G, Cheung WWL, Connors S, Denton F, Diongue-Niang A, Dodman D, Garschagen M, Geden O, Hayward B, Jones C, Péan C (2023) IPCC, 2023: Climate change 2023: synthesis report. In: Lee H, Romero J (eds) Contribution of working groups I, II and III to the sixth assessment report of the intergovernmental panel on climate change, IPCC, Geneva, Switzerland. (First), Intergovernmental Panel on Climate Change (IPCC). https://doi.org/10.59327/IPCC/AR6-9789291691647

Campbell BM, Beare DJ, Bennett EM, Hall-Spencer JM, Ingram JSI, Jaramillo F, Ortiz R, Ramankutty N, Sayer JA, Shindell D (2017) Agriculture production as a major driver of the Earth system exceeding planetary boundaries. Ecol Soc 22(4). https://www.jstor.org/stable/26798991

Chandra A, McNamara KE, Dargusch P (2017) The relevance of political ecology perspectives for smallholder climate-smart agriculture: a review. J Polit Ecol 24(1). https://doi.org/10.2458/v24i1.20969

Chandra A, McNamara KE, Dargusch P (2018) Climate-smart agriculture: perspectives and framings. Clim Policy 18(4):526–541. https://doi.org/10.1080/14693062.2017.1316968

Chataut G, Bhatta B, Joshi D, Subedi K, Kafle K (2023) Greenhouse gases emission from agricultural soil: a review. J Agric Food Res 11:100533. https://doi.org/10.1016/j.jafr.2023.100533

Díaz-Varela ER, Álvarez-Álvarez P, Roces-Díaz JV, Rodríguez-Morales B (2018) The contribution of chestnut orchard recovery projects for effective area-based conservation: two cases in Asturias (North-West Spain)—international Partnership for the Satoyama initiative. International Partnership for the Satoyama Initiative. https://satoyama-initiative.org/case_studies/the-contribution-of-chestnut-orchard-recovery-projects-for-effective-area-based-conservation-two-cases-in-asturias-north-west-spain/

Dinesh D, Aggarwal P, Khatri-Chhetri A, Rodríguez AML, Mungai C, Sebastian L, Zougmore RB (2017) The rise in climate-smart agriculture strategies, policies, partnerships and investments across the globe. Agric Dev 30. https://oar.icrisat.org/10017/

Dobermann A (2004) A critical assessment of the system of rice intensification (SRI). Agric Syst 79(3):261–281. https://doi.org/10.1016/S0308-521X(03)00087-8

FAO (2010) "Climate-smart" agriculture: policies, practices and financing for food security, adaptation and mitigation. FAO. https://www.fao.org/4/i1881e/i1881e00.htm

FAO (2021) Strategic framework 2022–31. FAO. https://openknowledge.fao.org/server/api/core/bitstreams/29404c26-c71d-4982-a899-77bdb2937eef/content

Gevaña DT, Garcia JE, Ruzol CD, Malabayabas FL, Grefalda LB, O'Brien E, Santos EP, Camacho LD (2021) Climate change resiliency through mangrove conservation: the case of Alitas farmers of Infanta, Philippines. In: Nishi M, Subramanian SM, Gupta H, Yoshino M, Takahashi Y, Miwa K, Takeda T (eds) Fostering transformative change for sustainability in the context of socio-ecological production landscapes and seascapes (SEPLS). Springer, pp 195–214. https://doi.org/10.1007/978-981-33-6761-6_11

Huyer S, Partey S (2020) Weathering the storm or storming the norms? Moving gender equality forward in climate-resilient agriculture. Clim Change 158(1):1–12. https://doi.org/10.1007/s10584-019-02612-5

IPBES (2024) Summary for policymakers of the thematic assessment report on the underlying causes of biodiversity loss and the determinants of transformative change and options for achieving the 2050 vision for biodiversity of the intergovernmental science-policy platform on biodiversity and ecosystem services. IPBES secretariat. https://zenodo.org/records/14513975

IPSI (2019) ABOUT IPSI. International Partnership for the Satoyama Initiative. https://satoyama-initiative.org/about/

IPSI (2020) The Satoyama initiative. International Partnership for the Satoyama Initiative. https://satoyama-initiative.org/concept/satoyama-initiative/

Karlsson L, Naess LO, Nightingale A, Thompson J (2018) 'Triple wins' or 'triple faults'? Analysing the equity implications of policy discourses on climate-smart agriculture (CSA). J Peasant Stud 45(1):150–174. https://doi.org/10.1080/03066150.2017.1351433

Li J, Xia E, Wang L, Yan K, Zhu L, Huang J (2022) Knowledge domain and emerging trends of climate-smart agriculture: a bibliometric study. Environ Sci Pollut Res 29(46):70360–70379. https://doi.org/10.1007/s11356-022-20796-9

Lipper L, Thornton P, Campbell BM, Baedeker T, Braimoh A, Bwalya M, Caron P, Cattaneo A, Garrity D, Henry K, Hottle R, Jackson L, Jarvis A, Kossam F, Mann W, McCarthy N, Meybeck A, Neufeldt H, Remington T, Torquebiau EF (2014) Climate-smart agriculture for food security. Nat Clim Chang 4(12):1068–1072. https://doi.org/10.1038/nclimate2437

Mafongoya PL, Ajayi OC (eds) (2017) Indigenous knowledge systems and climate change management in Africa. CTA

Mbawine JS, Dzekoto GE (2023) Community-based woodland restoration for livelihoods and sustainable wood fuel utilisation in the mole ecological landscape, Ghana. In: Nishi M, Subramanian SM (eds) Ecosystem restoration through managing socio-ecological production landscapes and seascapes (SEPLS). Springer Nature, pp 13–33. https://doi.org/10.1007/978-981-99-1292-6_2

Méndez-López ME, Cepeda-González MF, Rodríguez-Robayo KJ, Juárez-Téllez L, Rivera-De Velasco M, Peralta-Blanco RM, Chan-Chuc N, Serrano-Ysunza AA, Riveros-Cañas RA, Sánchez-Siordia OG, Proust S (2022) "To take care of the land means taking care of ourselves": local perceptions on human and environmental health in a high agro-biodiversity landscape in the Yucatan Peninsula. In: Nishi M, Subramanian SZ, Gupta H (eds), Biodiversity-health-sustainability nexus in socio-ecological production landscapes and seascapes (SEPLS). Springer Nature, pp 89–105. https://doi.org/10.1007/978-981-16-9893-4_5

Moore M-L, Riddell D, Vocisano D (2015) Scaling out, scaling up, scaling deep: strategies of non-profits in advancing systemic social innovation. J Corp Citiz 58:67–84. https://www.jstor.org/stable/jcorpciti.58.67

Pinto AD, Cenacchi N, Kwon H-Y, Koo J, Dunston S (2020) Climate smart agriculture and global food-crop production. PLoS ONE 15(4):e0231764. https://doi.org/10.1371/journal.pone.0231764

Schuelke-Leech B-A (2021) Disruptive technologies for a Green New Deal. Curr Opin Environ Sci Health 21:100245. https://doi.org/10.1016/j.coesh.2021.100245

Shahzad A, Ullah S, Dar AA, Sardar MF, Mehmood T, Tufail MA, Shakoor A, Haris M (2021) Nexus on climate change: agriculture and possible solution to cope future climate change stresses. Environ Sci Pollut Res 28(12):14211–14232. https://doi.org/10.1007/s11356-021-12649-8

Taylor M (2018) Climate-smart agriculture: what is it good for? J Peasant Stud 45(1):89–107. https://doi.org/10.1080/03066150.2017.1312355

Thakur AK, Uphoff NT (2017) How the system of rice intensification can contribute to climate-smart agriculture. Agron J 109(4):1163–1182. https://doi.org/10.2134/agronj2016.03.0162

The World Bank. (2024). *Climate-Smart Agriculture*. https://www.worldbank.org/en/topic/climate-smart-agriculture

Thomas RJ, O'Hare G, Coyle D (2023) Understanding technology acceptance in smart agriculture: a systematic review of empirical research in crop production. Technol Forecast Soc Chang 189:122374. https://doi.org/10.1016/j.techfore.2023.122374

UNDP (2017) COMDEKS Project: Iipumbu-ya-Tshilongo conservancy, Namibia. International Partnership for the Satoyama Initiative. https://satoyama-initiative.org/case_studies/comdeks-project-iipumbu-ya-tshilongo-conservancy-namibia/

UNU-IAS, Bioversity International, IGES, & UNDP (2014) Toolkit for the indicators of resilience in socio-ecological production landscapes and seascapes. https://hdl.handle.net/10568/104521

Vishnoi S, Goel RK (2024) Climate smart agriculture for sustainable productivity and healthy landscapes. Environ Sci Policy 151:103600. https://doi.org/10.1016/j.envsci.2023.103600

Wakweya RB (2023) Challenges and prospects of adopting climate-smart agricultural practices and technologies: Implications for food security. J Agric Food Res 14:100698. https://doi.org/10.1016/j.jafr.2023.100698

Wang C-W, Lee K-C, Huang H-H, Chen D-W (2023) Incorporating green energy into smart agriculture to create regional circular and ecological spheres (R-CES)—international partnership for the Satoyama initiative. International Partnership for the Satoyama Initiative. https://satoyama-initiative.org/case_studies/incorporating-green-energy-into-smart-agriculture-to-create-regional-circular-and-ecological-spheres-r-ces/

Wheeler T, von Braun J (2013) Climate change impacts on global food security. Science 341(6145):508–513. https://doi.org/10.1126/science.1239402

World Bank Group (2012) Climate-smart agriculture: a call to action [Working Paper]. World Bank Group. https://documents.worldbank.org/en/publication/documents-reports/documentdetail/992021468197391264/Climate-smart-agriculture-a-call-to-action

Yang Y, Tilman D, Jin Z, Smith P, Barrett CB, Zhu Y-G, Burney J, D'Odorico P, Fantke P, Fargione J, Finlay JC, Rulli MC, Sloat L, Jan van Groenigen K, West PC, Ziska L, Michalak AM, the Clim-Ag Team, Lobell DB (2024) Climate change exacerbates the environmental impacts of agriculture. Science 385(6713):eadn3747. https://doi.org/10.1126/science.adn3747

Zhao J, Liu D, Huang R (2023) A review of climate-smart agriculture: recent advancements, challenges, and future directions. Sustainability 15(4):4. https://doi.org/10.3390/su15043404

Novel Financial Tools and Disruptive Technologies

Green and Sustainable Municipal Bonds: Accelerating Infrastructure Investment for Equitable and Sustainable Cities

Wenjie Liu, Mahesti Okitasari, and Eric Lau

1 Introduction

Among the existing global greenhouse gas (GHG) emissions, those from energy, transportation, construction, and water infrastructure account for over 60% of total emissions (OECD 2018). Although many countries and regions have implemented regulations and policies to support green and climate-resilient infrastructure construction and encourage investors and businesses to move towards more sustainable practices, both developed and developing economies face the long-term challenge of insufficient investment in sustainable infrastructure (Cheshmehzangi et al. 2021; OECD 2024). The Organisation for Economic Co-operation and Development (OECD) estimates that approximately USD 6.9 trillion will need to be invested annually in new infrastructure before 2030 to achieve climate and development goals (OECD 2017a). This financing gap will keep increasing due to economic and financial shocks.

Within the framework of the Paris Agreement and in the slow uptake of carbon pricing schemes globally, bond markets become central to financing the transition to a global low-carbon economy (OECD 2017b). The first green bond was issued in 2007 with the AAA-rated issuance by the European Investment Bank, which was then followed by the World Bank. There are several types of green bonds, yet the most common is "use of proceeds" bond (e.g. Barclay Green Bond, Climate Awareness Bond) where proceeds raised by bond sale are earmarked for green projects (CBI 2021b). On the other hand, sustainability bonds enable capital-raising and investments where the proceeds will be exclusively applied to finance or refinance a combination of green and social projects (ICMA 2021). Sustainability bonds differ from sustainability-linked bonds where the performance of the security is linked to

W. Liu (✉) · M. Okitasari
United Nations University Institute for the Advanced Study of Sustainability, Tokyo, Japan

E. Lau
Beijing Circular Economy Strategies, Beijing, China

M. Okitasari et al. (eds.), *Equitable and Sustainable Cities*, Advances in 21st Century Human Settlements, https://doi.org/10.1007/978-981-92-2793-8_7

sustainability performance targets of the issuer instead of constraining the use of funds. An example of sustainability bonds issued in developing regions is Chile's Arauco sustainability bonds, starting in 2019.

Green and sustainability bonds are increasingly recognised as important instruments for bridging the finance gap and driving rapid growth in investments focussed on green and sustainable initiatives. By 31 December 2023, Climate Bonds had recorded cumulative issuance of GSS+ instruments (green, social, and sustainability bonds, along with sustainability-linked bonds) totalling USD 4.4 trillion (CBI 2023b). Green bonds remain the largest segment of the market—70 and 66% of overall issuances in developing and developed markets, respectively (World Bank 2024). However, social and sustainability bonds have experienced significant growth since 2020. Public sector issuances represent 33% of the total amount of GSS+, with sovereigns as the biggest segment (25 sovereigns by 2024), followed by government agencies and government development banks (World Bank 2024). Bonds issued by local governments have also increased moderately, but they account for only a small portion of the market—just 5.6% in 2023 (CBI 2023b). The growth of the market has been uneven, with developed markets issuing more than developing markets, and activity is geographically concentrated in China, the United States (US), and several European nations. Up to the second quarter of 2024, the share of the developing market within GSS+ is only 16% of the total amount issued, yet it demonstrates a stronger preference for sustainability bonds compared to the developed market (World Bank 2024). Among 36 developing market sovereigns, Chile remains the largest GSS+ bond issuer with USD 48 billion, followed by Thailand, Mexico, Peru, and Indonesia with at least USD 10 billion in bonds issued (World Bank 2024). Figures 1 and 2 illustrate GSS+ market growth and geographical distribution between 2014 and 2023.

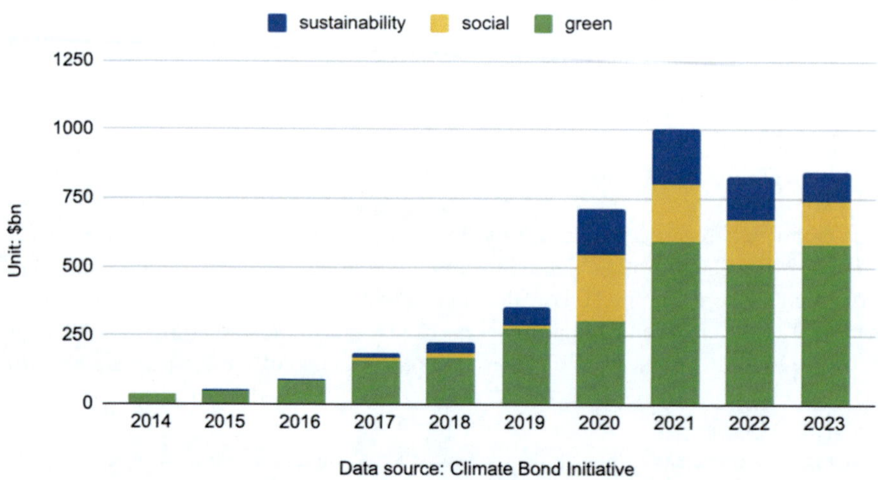

Fig. 1 Green, social, and sustainability bonds market growth by year

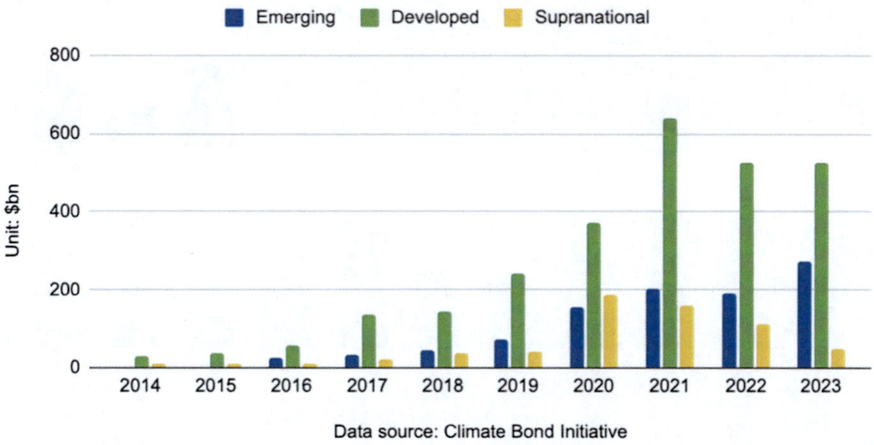

Fig. 2 Distribution across emerging, developed, and supranational markets

Transition bonds are a type of bond instrument designed to support entities or projects in high-emission industries as they transition towards a more sustainable and low-carbon future. While it is often referred to as a subset of green bonds (e.g. Securities and Exchange Board of India considers transition bonds as a sub-category of their green debt securities), it is more apt to categorise transition bonds as between green bonds and sustainability-linked bonds—transition bonds fund the issuers to become less carbon-intensive with proceeds used for specific decarbonisation-focussed projects though not necessarily to become completely green (Shrimali 2021). Given the lack of definitions and eligibility criteria, lock-in risks are highly present in transition bonds (OECD 2023). As of 2023, a cumulative amount of USD 11.3 billion in transition bonds has been issued (Huan et al. 2020) with the potential to see significant development in Asia (ICMA 2024). The Japanese government has been a leader in promoting the use of transition bonds (Ministry of Finance 2024) with USD 5.4 billion issued in 2024 as 10-year Japan Climate Transition Bonds (ICMA 2024), although direct issuance by local governments remains an emerging practice.

Regional and local governments are important investors in green and climate-resilient infrastructure (OECD 2024). In the expanding cities of the developing world, such infrastructures are often very poor or lacking due to a combination of planning and investment issues (Cheshmehzangi et al. 2021). In recent years, the mechanisms of international and domestic institutions have grown to address the potential of optimising scarce public finance and spending in ways that generate more participation from the private sector to bridge the gap (UN-Habitat 2021). Municipal bonds, traditionally used to finance public infrastructure projects, are undergoing a transformative shift to finance sustainable development. Green and sustainable municipal bonds are seen as a natural fit for green and climate-resilient infrastructure projects (CBI 2021a). Green municipal bonds specifically earmark funds for projects with environmental benefits, such as renewable energy installations, energy-efficient buildings,

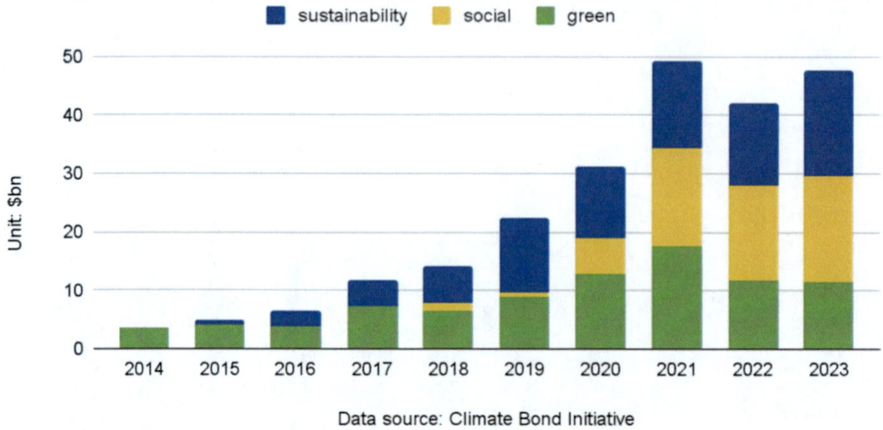

Fig. 3 Green, social, and sustainability bonds issued by local governments. *Source* Authors

and green transportation initiatives (Yun 2020). Sustainable municipal bonds, on the other hand, mainly utilise financial instruments such as social bonds, transition bonds, sustainability bonds, and sustainability-linked bonds to raise funds aimed at supporting sustainable development-related goals and purposes (Liu and Liu 2024). These innovative financial instruments hold promising potential for addressing the dual challenge of urbanisation and climate change (Fig. 3).

Issuing green municipal bonds offers numerous benefits. Not only do they come with lower financing costs, but they can also effectively bridge fiscal revenue gaps, quickly raise substantial funds for infrastructure development and improve urban environments (Gorelick et al. 2024; Suzuki et al. 2020; Wijaya et al. 2021). Additionally, issuing these bonds can enhance the reputation and transparency of the issuing government and attract support from investment institutions that prefer sustainable investments (C40 2015; Kahagalle and Lindeman-Jones 2021; True 2023). Currently, green municipal bonds have accumulated substantial practical experience in some developed countries. European cities such as Gothenburg, London, Paris, and Stockholm, and US cities such as Los Angeles, San Francisco, St Paul, and Washington, DC, are pioneer issuers of green municipal bonds in developed markets (CBI et al. n.d.). Since 2013, green municipal bonds have proliferated in US cities (Karpf and Mandel 2017) and are gradually emerging in the rest of the global North and South (Herrera 2024). Johannesburg in South Africa is the first one in the developing market with issuance in 2014, followed by Mexico City in 2016 as the first city in Latin America to issue green municipal bonds (OECD 2017a, b). Green municipal bonds can be certified by independent bodies and certified bonds typically have more favourable pricing (Gorelick et al. 2024). Examples of cities in the global South receiving CBI certification are South Africa's Cape Town (July 2017) and India's Vadodara Municipal Corporation (March 2024). Both cities use the proceeds to finance water infrastructure projects. The adoption of green municipal bonds in cities in developing nations remains less prevalent. These cities encounter various barriers, including a lack of standardisation, limited investor awareness, concerns about market size and liquidity, data and reporting challenges, and higher administrative and verification costs (Gorelick et al. 2024).

Green bonds have attracted substantial attention from investors, policymakers, and researchers, creating a vast body of literature that explores their role in sustainable development, climate finance, and the broader financial market. The discussion has been centred around the definition (ICMA n.d.) and evolution of standards (Packer and Ehlers 2017), market growth and trends (CBI 2024), financial performance (Flammer 2021; Zerbib 2019) and investment strategies, and environmental impact. As a specific type of green bond, green municipal bonds have garnered significant attention as the market has grown in recent years for cities to unlock more capital for more sustainable infrastructure projects. Despite the increasing interest in green municipal bonds as a tool for financing sustainable infrastructure, several research gaps remain.

One significant research gap is the lack of standardised criteria and frameworks for green and sustainable municipal bonds. While the International Capital Market Association (ICMA) provides Green Bond Principles, these guidelines are voluntary and not uniformly applied across different jurisdictions (Shishlov et al. 2016). This has resulted in inconsistent definitions of what constitutes a "green" project, which in turn raises concerns about greenwashing (European Parliament 2022; Ghaemi et al. 2024). Existing literature largely focusses on the benefits of green municipal bonds (Brand and Steinbrecher 2019; Dembele et al. 2021), origins and financial constructions (Herrera 2024), and practical challenges (Gorelick et al. 2024), but there is a need for more in-depth studies on how variations in certification processes, third-party verifications, and reporting practices impact investor confidence and bond performance.

Although green and sustainable municipal bonds are designed to finance projects with positive environmental outcomes, there is limited empirical evidence on the long-term environmental and financial impacts of these projects. Existing studies focus more on the economic benefits, such as reduced borrowing costs through the "greenium," but less is known about how these projects contribute to long-term climate resilience, carbon reduction, and environmental sustainability (Karpf and Mandel 2017). There is an urgency to understand the balance between the benefits of green and sustainable municipal bonds and their financial implications, especially when compared to traditional bonds, which represent the largest fraction of the financial market (Gorelick et al. 2024). Additionally, more research is needed to assess how municipalities track and report on the sustainability outcomes of the projects financed by these bonds, particularly over extended periods.

The green and sustainable municipal bond market is a relatively young and evolving field, with much still to be understood about its long-term impacts, inclusivity, and role in addressing climate change at the local level. Given the ambiguity of green and sustainable municipal bonds' impact on equitable and sustainable cities, this chapter explores their role and contributions as a strategic financial instrument to expedite urban transformation through accelerated infrastructure investment.

2 Methodology

This research employs a case study approach to analyse the development of green and sustainable municipal bonds in the global North and South. Following (Flyvbjerg 2011) in maximising the utility of information from small samples and single cases, an information-oriented approach is employed to select cases. As the issuance of green and sustainable municipal bonds is still emerging and remains limited within the bond market, four case studies from Canada, China, Hong Kong, and the USA are selected on the basis of expectations about their information content relevant to the development and implementation of green and sustainable municipal bonds, which can offer lessons learned to understand their role and contributions—and limitations—in delivering sustainable infrastructure at the local level. Additionally, the selection of case studies considers the market status (developing vs developed, volume, and trends) and the country's history and recent development of municipal and green and sustainable municipal bond issuances to provide diverse characteristics of cases. Canadian and US cities both have a long history of municipal bonds, and US cities in particular are among the first ones to issue green municipal bonds (Massachusetts in 2013). In 2023, Hong Kong was the fourth largest source of aligned government green bonds behind the United Kingdom, Germany, and Italy and historically a diverse debt market (CBI 2023a). China's case is expected to provide a different perspective from that of a bank-oriented bond market compared to a market-oriented bond market such as Canada and the USA. One unique feature of China is that local governments formulate their own plans to develop green finance, and the volume of the local government bond market is massive (Huan et al. 2020).

The case studies provide in-depth evaluations of green bond issuance frameworks and alignment with international standards, revenue generation models, and post-issuance disclosures, especially on environmental benefits. Data is gathered from a range of sources, including bond rating reports, policy documents, certification reports, and other publicly available information. Each case explores the unique aspects of policy frameworks and standards, use of proceeds, revenue mechanisms, disclosures, and green certification methods used by different issuers. Special emphasis is placed on how these bonds align with sustainable infrastructure investment and uphold high environmental standards. The key takeaways from these case studies are synthesised to offer valuable insights for policymakers, financial institutions, and urban practitioners on leveraging innovative financial instruments to align urban infrastructure investments with sustainable development goals.

3 Case Study

3.1 Case Study 1: Robust Policy Support and Revenue Safeguards

China's green bond market started in 2016 and has developed rapidly, leading issuance for the second consecutive year in 2023 with a total of USD 131.3 billion (RMB 0.94 trillion) labelled green bonds in domestic and overseas markets (CBI 2024). The country's views on tackling climate change as an important angle in its long-term strategic economic and social development objective have underpinned the proliferation of labelled themes attached to green, transition, social, and sustainability in the bond market (CBI 2024). With the development of the green bond market, some provincial governments in China have started exploring the "greening" of local government bonds to attract private capital and finance infrastructure projects with significant ecological and environmental benefits. According to the Green Bond Endorsed Project Catalogue (2021 Edition), green bonds refer to securities issued through legal procedures where the funds raised are specifically allocated to support green industries, green projects, or green economic activities that meet the defined conditions. These activities are primarily categorised into six major areas: energy conservation and environmental protection industries, clean production industries, clean energy industries, ecological and environmental industries, green infrastructure upgrades, and green services (People's Bank of China 2021). Representative examples include the Jiangxi Provincial Government's issuance of the Jiangxi Ganjiang New District Green Municipal Special Bond in 2019 and the Guangdong Provincial Government's issuance of the Pearl River Delta Water Resources Allocation Project Special Bond in 2020, hereafter referred to as the "Jiangxi Government Green Special Bond" and the "Guangdong Government Green Special Bond," respectively (Table 1).

Both regions where these green special bond-supported projects are located—Jiangxi's Ganjiang New District and Guangzhou in Guangdong Province—are Green Finance Reform and Innovation Pilot Zones. The provincial governments in these two areas were pioneers in experimenting with green special bonds, and both were oversubscribed after being issued and listed. The Jiangxi Government's green special bond, China's first 30-year green municipal special bond, was oversubscribed more than 12 times after its issuance, injecting confidence into the market (Yun 2020). The two green special bond projects share the following highlights.

3.1.1 Advanced and Demonstrative Nature of Green Projects

The "Guangdong Government Green Special Bond" is specifically used to support the construction of a major provincial water conservation project—the Pearl River Delta Water Resources Allocation Project. This project is one of the key water conservation and supply projects approved by the State Council, and it is the largest, longest, and

Table 1 Green local government bonds in Ganzhou New District and Guangzhou City. *Source* Authors

Bond	Issuer	Scale (million USD)	Interest rate (%)	Term (year)	Funded project	Repayment source	Environmental KPIs
2020 Guangdong Province Special Bond Sixty Issues	Guangdong Provincial Government	368.24	2.88	10	Pearl River Delta Water Resources Allocation Project	Water fee	Save energy and water resources and calculate the carbon emissions, energy, and water resources that can be reduced after the project is completed (compared to direct buried pipelines)
2019 Jiangxi Provincial Government Special Bond Seventh Issue	Jiangxi Provincial Government	40.9	4.11	30	Ganjiang New District Rule Lake New City No. 1 Pipe Gallery and Smart Pipe Gallery Project	Corridor entry fee, utility tunnel maintenance fee, comprehensive development income, and financial subsidy income of the pipe gallery project	Improve water supply in the eastern region of the Pearl River Delta, improve urban water security, increase ecological water use, and improve the environment

Note Information collected and analysed by the authors based on the bond rating reports

most extensive water transfer project in Guangdong's history. The goal of the project is to divert water from the Xijiang River system to the eastern part of the Pearl River Delta, addressing the water shortage for residential and industrial use in Dongguan, Guangzhou, and Shenzhen, while also providing emergency backup water supply for regions such as Foshan, Guangzhou, and the Hong Kong Special Administrative Region.

The "Jiangxi Government Green Special Bond" is specifically dedicated to supporting the construction of the Ganjiang New District comprehensive utility tunnel and smart utility tunnel projects. Urban underground utility tunnels are key infrastructure projects that centralise the installation of city pipelines for water supply, rainwater, sewage, recycled water, natural gas, heating, electricity, and telecommunications. These tunnels enable the scientific and efficient use of urban underground space, enhance a city's disaster prevention and emergency response capabilities, and strengthen urban resilience.

3.1.2 Disclosure of Environmental Benefits Before Bond Issuance

Both green special bonds underwent independent evaluations and certifications as green bonds prior to issuance. According to the pre-issuance independent evaluation report for the "Jiangxi Government Green Special Bond," the project falls under the category of energy-saving urban infrastructure (ChinaBond n.d.a). It avoids material consumption, energy use, and carbon emissions associated with the maintenance and management of directly buried pipelines and the secondary excavation of roads. It also prevents gas and water leakage from directly buried pipelines. Before issuance, the bond calculated and disclosed the expected savings in energy and water resources, as well as reductions in carbon emissions upon the completion of the project.

The pre-issuance disclosure documents for the "Guangdong Government Green Special Bond" provided detailed descriptions of the project's environmental and ecological benefits, as well as its resulting economic and social benefits (ChinaBond n.d.b). The project is expected to enhance the city's water resource capacity, ensure water supply security, improve the water environment, and promote the economic development of urban clusters. By diverting water from the west to the eastern water-scarce areas, the water supply–demand imbalance in Guangdong Province will be alleviated, while the ecological water flow previously displaced in the river channels will be restored, improving the river environment. After addressing the bottleneck of insufficient water resources, the industrial growth and livability of the Pearl River Delta urban cluster will be improved.

3.1.3 Robust Revenues After Operation Provide a Guarantee for Debt Repayment

The project supported by the "Guangdong Government Green Special Bond" is government-led, adopting a corporate management strategy. The source of debt repayment comes from water fees collected after the project is operational, with the project company Guangdong Yuehai Pearl River Delta Water Supply Co., Ltd. charging fees from Guangzhou, Shenzhen, and Dongguan. The project was implemented by Guangdong Yuehai Pearl River Delta Water Supply Co., Ltd. Before the bond issuance, the Guangdong Provincial Water Resources Department publicly disclosed the project details and certified the financing estimates. According to the certification report, the project's debt coverage ratio during the bond's term is 1.29, indicating sufficient funds to cover principal and interest repayments (ChinaBond n.d.b).

The underground utility tunnel project supported by the "Jiangxi Government Green Special Bond" is expected to generate revenue from four main sources: utility tunnel usage fees, pipeline maintenance fees, comprehensive development income, and fiscal subsidy income. Among these, utility tunnel usage fees and pipeline maintenance fees are the primary sources of project revenue. These fees are set based on guidelines from the National Development and Reform Commission (NDRC) and the Jiangxi Provincial Government, taking into account normal operating costs, taxes, interest rates, and reasonable operating profits for the utility tunnels. In the early stages of operation, necessary fiscal subsidies will be provided to compensate for any fee shortfalls. In the long term, the cash flow generated from the four revenue streams will be able to cover the debt principal and interest. The two tunnels included in the project have debt coverage ratios of 1.22 and 1.59 during the bond term, and full lifecycle debt coverage ratios of 1.10 and 1.12, respectively (ChinaBond n.d.a).

Both bonds successfully obtained green labelling and committed to allocating the raised funds exclusively for green projects, supported by favourable policy conditions and a robust foundation of project reserves. The expected revenues from the projects provide a safeguard for the interest repayments, thereby reducing excessive reliance on local government fiscal income.

3.2 *Case Study 2: Aligning with International Standards to Establish Local Norms*

China's offshore green bond market has also begun experimenting with green municipal bonds, drawing on the experience of the Hong Kong government. This market is characterised by large-scale issuances, multiple currencies, long maturities, support for diverse projects, and alignment with international standards. So far, the Hong Kong government has issued green bonds in multiple currencies, including a record-setting 30-year, USD 500 million green bonds issued in 2021 (Hong Kong Monetary Authority 2021). In 2022, the Hong Kong government issued its first green retail bond, with a scale of HKD 20 billion (USD 2.57 billion) (The Hong Kong Special Administrative Region Government 2022). In addition to innovations in bond types and scale, Hong Kong's green bond issuance plan is aligned with mainstream international Green Bond Principles, demonstrating advanced practices in project selection, issuance plan supervision, and post-issuance information disclosure. Several takeaways from the Hong Kong case are discussed below.

3.2.1 Establishing Green Bond Issuance Standards in Line with International Standards

The Hong Kong government first published its "Green Bond Framework" in 2019, which aligns with the ICMA "Green Bond Principles" to guide and regulate the government's green bond issuance plan. In 2022, following the updates to ICMA's Green Bond Principles, the Hong Kong government also updated its "Green Bond Framework." The framework specifies that the proceeds from government green bonds can be used to finance or refinance projects in nine categories: renewable energy, energy efficiency, pollution prevention, waste management and resource recovery, water and wastewater management, conservation and biodiversity, clean transportation, green buildings, and climate change adaptation. The framework not only outlines the eligible project categories but also provides technical benchmarks, including defining project goals and scope, setting carbon emission intensity and energy efficiency requirements for certain projects, and establishing certification requirements. For instance, the framework mandates that waste-to-energy projects in the waste management, and resource recovery category must achieve at least a 25% waste-to-energy conversion efficiency ratio and maintain a carbon dioxide equivalent emission intensity of less than 100 g/kWh. The framework also includes exclusionary clauses, stipulating that funds cannot be used for projects related to fossil fuel power generation or improving the efficiency of fossil fuel power generation (The Government of the Hong Kong Special Administrative Region 2022).

3.2.2 Establishment of a Cross-Departmental Green Bond Committee

The Hong Kong government has established a "Green Bond Program Steering Committee" to oversee the issuance of government green bonds, allocate funds, review project eligibility, and report on progress. This steering committee is chaired by the Financial Secretary, with members including leaders from the Financial Services and the Treasury Bureau, the Environment and Ecology Bureau, and the Hong Kong Monetary Authority. The Financial Secretary sets borrowing limits in the budget, creating room for the issuance of government green bonds. The Hong Kong Monetary Authority is responsible for implementing the green bond issuance plan, and the funds raised are transferred to the government's Capital Works Reserve Fund, managed by the Financial Services and the Treasury Bureau.

3.2.3 Regular Disclosure of Fund Usage and Environmental Benefits of Projects

Each year, the Financial Services and the Treasury Bureau publishes a "Green Bond Report" detailing the allocation of raised funds, project progress, and the environmental benefits achieved by the projects. A qualified third party is commissioned to verify the content of the report. For the nine categories of projects that meet the "Green Bond Framework" requirements, the Hong Kong government quantifies the environmental benefits based on ICMA's recommended metrics and discloses the information accordingly.

3.3 *Case Study 3: Multiple Green Certification Methods*

The bulk of green municipal bond issuance is concentrated in Europe and the USA. Compared to the nascent green municipal bond markets in Asia, the USA has accumulated nearly 10 years of experience in issuing green municipal bonds, with a larger market, a greater variety of issuers, and more appeal to individual investors. Issuers of green municipal bonds in the USA include state governments, as well as various levels of local governments and government-designated agencies. The projects supported are mainly in the areas of water resources, sewer projects, public power, green transportation, and green buildings (e.g. energy-efficient school buildings and dormitories) (Baker et al. 2022; Forsgren 2016). US green municipal bonds also have around 1.66 years longer maturities and are less likely to be bank eligible (fixed costs of green status make green municipal bonds on average larger than ordinary bonds or part of larger bond issues) compared to ordinary municipal bonds (Baker et al. 2022).

In terms of bond types, US green municipal bonds include not only project bonds, where the repayment source comes from the project's revenue but also general obligation bonds and revenue bonds, with varying levels of recourse and repayment sources (Zeng and Wu 2019). General obligation bonds give full recourse to the issuer, while green revenue bonds provide recourse to the issuer's collected taxes and fees. In addition, issuers can bundle multiple projects into a single green municipal bond, enabling simultaneous financing for several green projects.

In order to highlight the "green" attributes of municipal bonds, issuers disclose the basis for the green designation in the Official Statement (OS) documents provided to investors during bond issuance. They also continue to disclose the use of funds and project progress after issuance. Currently, there are various forms of green certification for US municipal bonds, and some bonds labelled as "green" involve external evaluations to strengthen their green credentials. Given the characteristics of US green municipal bonds described before, the following highlights lessons learned on external reviews, verification, and ratings in New York, Oregon, San Francisco, and Washington, DC.

3.3.1 Introduction of Independent Second-Party Opinions

In 2014, the Washington DC Water and Sewer Authority issued a revenue bond for water resources, marking the first US green municipal bond to receive an independent Second Party Opinion (SPO). The proceeds from this bond were used to construct urban stormwater and wastewater collection and treatment facilities. The independent evaluation agency assessed the issuer's environmental, social, and governance (ESG) performance, the environmental and social benefits of the project, and the information disclosure, thereby recognising the "green" labelling of this bond.

3.3.2 Green Bond Verification

In 2022, the New York Power Authority (NYPA) issued a green project bond to upgrade regional power grids. After verification by an external agency, this bond was deemed to comply with NYPA's previously released "Green Bond Framework" and was therefore labelled as "green." NYPA published its "Green Bond Framework" in 2020, committing to issuing green bonds to support green project construction and help achieve its goals of developing clean energy and combating climate change. This framework received a second-party opinion, affirming its alignment with the ICMA's Green Bond Principles.

3.3.3 Green Bond Certification

Since 2016, the San Francisco Public Utilities Commission (SFPUC) has issued several green revenue bonds to finance water resource projects, wastewater treatment projects, and clean energy transmission projects. Many of these bonds have been certified by the Climate Bonds Initiative (CBI). For instance, a bond issued in 2016 funded the construction of regional waterworks, and another bond issued in 2021 supported the construction of stormwater and wastewater collection and treatment facilities across San Francisco.

In addition to globally recognised green bond certifications, there are also local certification mechanisms in the USA. For example, the municipal bond insurance agency Build America Mutual (BAM) provides green bond certification for municipal bonds. BAM provided a repayment guarantee for a green general obligation bond issued by a city government in Oregon in 2021, evaluating the bond's compliance with ICMA's Green Bond Principles and labelling it as a BAM "GreenStar" bond. BAM's "GreenStar" certification has been awarded to more than 150 municipal bonds, with a total funding scale exceeding USD 2 billion (BAM n.d.).

3.3.4 Green Bond Rating

In 2021, the San Francisco Municipal Transportation Agency issued a green revenue bond aimed at improving traffic efficiency through the repair and construction of road facilities, with some of the facilities designed and built according to green building standards. The green attributes of this bond were recognised by a rating from S&P.

These various forms of green certification for municipal bonds primarily focus on issuance, aiming to verify the allocation of bond proceeds, fund management, and transparency, thereby enhancing bond visibility. More importantly, issuers must not only disclose this information before issuance but also conduct comprehensive, transparent disclosures of the actual ecological and environmental benefits generated by the bonds during their lifecycle, continuously enhancing the influence of green municipal bonds and increasing their attractiveness to investors.

3.4 *Case Study 4: Supporting Urban Development from Green to Sustainable*

With the ambitious goal of creating a "metropolitan city focussed on people, environment, and opportunities for living, working, and thriving," as well as city development strategies such as the "Greenest City 2020 Action Plan," "Transportation 2040," "Vancouver Economic Action Strategy," and the "Healthy City Strategy," Vancouver released a Green Bond Framework in 2018. This initiative aimed to support the city in achieving 100% renewable energy by 2050 and reducing greenhouse gas emissions by 80% compared to the 2007 level. Vancouver became the third Canadian city, following Ottawa and Toronto, to issue a green bond framework.

Initially, the framework included seven categories of green bond projects: renewable energy, energy efficiency, green buildings, clean transportation, pollution prevention and control, sustainable water and wastewater management, and environmentally sustainable natural resource management activities. These categories applied to green bonds issued after 30 June 30 2018. Vancouver set the green bond issuance guidelines in line with the four core components of the ICMA Green Bond Principles, which include the use of proceeds, project evaluation and selection, fund management, and reporting.

On 11 September 2018, Vancouver issued its first green bond, totalling CAD 85 million (USD 59 million) with a coupon rate of 3.10%. The principal and interest were paid from the city's general debt servicing fund (City of Vancouver 2018, n.d.b). To date, the initiative has financed several projects, including the redevelopment of Roddan Lodge affordable housing, following green building principles, and the reconstruction of stormwater and water supply pipelines to increase flood resilience.

In 2020, Vancouver replaced its Green Bond Framework with a Sustainable Development Bond Framework, integrating sustainability into its financing activities and promoting a transition towards a low-carbon, socially responsible, and climate-adaptive future. Vancouver's Sustainable Development Framework aligns with ICMA's Green Bond Principles, Social Bond Principles, and Sustainability Bond Guidelines across its four core components, ensuring robust standards for bond ratings, information disclosure, and investor engagement. The scope of eligible project types expanded to 11 categories, adding climate change adaptation and three social categories (City of Vancouver n.d.a). Key takeaways from Vancouver's Green and Sustainable Municipal Bond Issuance are as follows.

3.4.1 Closely Linked to Urban Sustainability Goals

From the issuance of a Green Bond Framework in 2018 that led to several green bonds being issued to the introduction of the Sustainable Development Bond Framework in 2020, Vancouver's municipal bond issuance not only reflects environmental benefits but also aligns with the city's overall planning. It provides unified, systematic support for the city's sustainability goals, effectively promoting economic, environmental, and social objectives.

3.4.2 Alignment with International Standards to Enhance Transparency

In developing the Green Bond Framework in 2018, Vancouver adopted international standards such as the ICMA's Green Bond Principles, improving the quality and feasibility of the city's standards. Drawing on international norms also boosts market transparency and comparability, attracting more investors to participate in municipal financing. Each year, Vancouver City updates reports on bond issuance, fund utilisation, and project progress on its government website under the "Investor Relations" page. These reports provide detailed information about bond issuance, execution, and use of funds, and they are accompanied by other relevant documents such as rating reports and Second Party Opinion (SPO) reports.

3.4.3 Rigorous Independent External Review

SPO is a common method of external review for green bonds, and Vancouver employs this approach, engaging globally recognised ESG rating agency Sustainalytics for the external assessment of its green and sustainable bond issuances. Sustainalytics evaluates the bond framework based on fund use, project selection, fund management, and reporting. It considers Vancouver's Sustainable Development Bond Framework to be "credible and impactful" and aligned with the 2018 Sustainability Bond Guidelines. Moreover, it recognises that the eligible project categories will generate positive environmental or social impacts and contribute to the achievement of the United Nations Sustainable Development Goals (SDGs).

Independent external reviews are crucial for enhancing the transparency of issuers' disclosures, and they significantly boost investor confidence in the issuer's claims and commitments. Credible second-party opinions like this provide robust support for Vancouver as it continues to issue sustainable development bonds to finance its low-carbon, socially responsible, and climate-resilient future.

Vancouver's experience in developing and implementing standards for green and sustainable bonds offers valuable lessons. Specifically, Vancouver's integration of urban development goals, infrastructure, and economic and financial policies into its municipal bond planning can help improve city infrastructure while simultaneously enhancing social welfare.

4 Conclusion and Policy Suggestions

This chapter focusses on green and sustainable municipal bonds, specifically exploring their role and contributions as a strategic financial instrument to expedite urban transformation through accelerated infrastructure investment. Amid mounting concern about the ambiguity of green municipal bonds due to the lack of standardised criteria and frameworks, this study investigates the extent and variety of the unique aspects of policy frameworks and standards, use of proceeds, revenue mechanisms, disclosures, and green certification methods used by different issuers. The case studies reveal the critical role that green and sustainable municipal bonds play in promoting sustainable infrastructure development across different regions.

The analysis reveals robust policy frameworks and green certifications, such as those aligned with ICMA principles (e.g. Green Bond Principles, Sustainable Bond Guidelines), significantly enhance market confidence and transparency, as demonstrated in the cases of Jiangxi and Guangdong in China, and Vancouver in Canada. These frameworks provide clear guidelines for project selection, management of funds, and environmental impact reporting, thereby attracting investors and reducing risk. It is worth noting that alignment with ICMA's Green Bond Principles is a common feature in the green municipal bond landscape. Cities like Johannesburg and Paris also follow similar principles, particularly in identifying projects.

Existing experience in pioneering cities shows that green municipal bonds can be successfully used for large-volume projects and enable higher investment. Our case studies further highlight that ensuring robust revenues after operation is key for debt repayment. With the risk of fiscal unsustainability at the municipal level highly prevalent, revenue safeguards through diversified repayment sources, such as water fees and utility usage fees, are needed to provide a strong financial foundation for green projects, ensuring the sustainability of debt repayment. Projects supported by green bonds are not only environmentally beneficial but also financially viable, as evidenced by high debt coverage ratios.

Washington, DC and Vancouver's experience shows that independent external reviews (e.g. SPOs) and third-party certifications (e.g. CBI certifications in the US) enhance credibility, transparency, and accountability. This builds investor trust, increasing the attractiveness of green bonds. In addition to independent evaluators and third-party certification, lessons learned from case studies show that alignment with local sustainable goals is necessary to avoid overlooking critical structural challenges and trade-offs related to the projects. Alignment with local sustainability goals provides a unified, systematic financial mechanism that supports the city's strategic sustainability initiatives by linking bond issuance with these sustainability targets that balance environmental protection with economic growth and social well-being. The alignment of the bond with the city's sustainable development or climate strategies can help reassure investors that their funds are utilised towards long-term and coordinated ends. Table 2 provides a comparative summary of the four case studies observed.

Table 2 Comparison of cases

Compared items	Case 1: Guangdong & Jiangxi, China	Case 2: Hong Kong	Case 3: Multiple Cities, US	Case 4: Vancouver, Canada
The government has developed policy and financial frameworks to support the issuance of green and sustainability bonds	Yes	Yes	N/A	Yes
The taxonomy applied to bond issuance aligns with ICMA or other international standards	N/A	Yes	Yes	Yes
The government has established a dedicated department to manage green municipal bonds	N/A	Yes	N/A	Yes
The bond has disclosed environmental KPIs	Yes	Yes	Yes	Yes
The bond has been verified by a third party or has undergone an external review	Yes	Yes	Yes	Yes
The bond aligns with the city's SDG goals	N/A	N/A	N/A	Yes

These case studies also provide a roadmap for other local governments seeking to leverage green bonds for sustainable development, offering lessons on best practices in policy design, financial structuring, and environmental accountability. It is essential for policymakers, practitioners, and researchers to reassess the role and value proposition of green and sustainable municipal bonds in the context of these considerations. First, improving regulations and related standards for green and sustainable municipal bonds and further clarifying the definition of green and sustainable bonds should be the initial steps taken in embarking on green municipal bond issuance. It is, therefore, important to incorporate considerations for climate adaptation and sustainable development and define the scope of projects supported, the use of raised funds, issuers, and pre- and post-issuance disclosure requirements.

Second, establishing inter-departmental cooperation and coordination mechanisms responsible for supervising the issuance of government green bonds, fund allocation, project eligibility review, and reporting is necessary to reduce the administrative burden and costs associated with green municipal bonds. The issuance process could be streamlined by integrating green and sustainable bonds into central and local budget planning and demand. Additionally, cities should prioritise improving the management of the entire bond project chain to avoid "emphasising application but neglecting management."

Third, building and maintaining a high-quality repository of green and sustainable projects is key to building an issuer's reputation. This indicates the need to develop criteria for selecting projects for green and sustainable municipal bonds and involve third-party institutions to provide feasibility studies or certification support for green projects. It is recommended that 100% of funds raised go towards green municipal bond projects, with social responsibility projects focussing on affordable basic infrastructure, essential services, food security, and sustainable food systems.

Fourth, improving the accounting and evaluation of environmental benefits and sustainable development performance of projects can help ease the monitoring process, raise the bar of urban development, and move the issuer a step in the direction of a more transparent mode of financing. Design appropriate performance evaluation methods based on different bond types to quantitatively assess the contribution of raised funds, ensuring that project investments lead to positive and significant environmental and sustainable development benefits. It is, thus, indispensable for cities to regularly disclose project progress and fund usage. This means ensuring that relevant disclosures are made before and after issuance, monitoring the use of raised funds, and enhancing the measurement and reporting of environmental and social benefits. Where feasible, adopt quantitative indicators and disclose fund usage at least annually through reports or website announcements, ensuring that the reports are clear, transparent, and verifiable.

Lastly, the central government could and should encourage local governments to issue green and sustainable municipal bonds by providing financial incentives. The central government can share the issuance costs by offering issuance subsidies, reducing financing costs, and enhancing credit through mechanisms such as guarantee funds, bond–loan combinations, and dedicated construction fund combinations, thereby reducing investment risks. This builds on the investors' confidence and interest in green municipal bonds.

Declaration of Competing Interests This study was part of Wenjie Liu and Eric Lau's work in an environmental non-governmental organisation. Mahesti Okitasari's contribution was made possible with the support from the Grant for Global Sustainability (GGS) of the Ministry of Education, Culture, Sports, Science, and Technology (MEXT) and the Ministry of Environment, Japan. There are no additional activities to disclose. The authors declare that they have no known competing financial interests or personal relationships that could have appeared to influence the work reported in this manuscript.

References

Baker M, Bergstresser D, Serafeim G, Wurgler J (2022) The pricing and ownership of US green bonds. Annu Rev Financ Econ 14(1):415–437. https://doi.org/10.1146/annurev-financial-111620-014802

BAM (n.d.) BAM credit profiles [Dataset]. https://bambonds.com/credit-profiles/

Brand S, Steinbrecher J (2019) Green bonds—a sustainable alternative for municipal infrastructure finance? https://www.kfw.de/PDF/Download-Center/Konzernthemen/Research/PDF-Dokumente-Fokus-Volkswirtschaft/Fokus-englische-Dateien/Fokus-2019-EN/Fokus-No.-245-March-2019-Green-Bonds.pdf

C40 (2015) How to issue a Green Muni Bond: The Green Muni Bonds Playbook. https://www.c40knowledgehub.org/s/article/How-to-issue-a-Green-Muni-Bond-The-Green-Muni-Bonds-Playbook?language=en_US

CBI (2021a) Green infrastructure investment opportunities in the Greater Bay Area. https://www.climatebonds.net/files/reports/cbi_gba_giio_chinese.pdf

CBI (2021b) Post-issuance reporting in the green bond market. https://www.climatebonds.net/files/reports/cbi_post_issuance_2021_02g.pdf
CBI (2023a) Hong Kong Green Market Briefing 2023. https://www.climatebonds.net/files/reports/cbi_hk_sotm_23_02e.pdf
CBI (2023b) The climate bonds interactive data platform [Dataset]. https://www.climatebonds.net/market/data/
CBI (2024) China sustainable debt-state of the market report 2023. https://www.climatebonds.net/files/reports/china_sustainable_debt_state_of_the_market_report_2023.pdf
CBI, CDP, C40 Cities, ICLEI, & ICMIF (n.d.) How to issue a green city bond. https://www.climatebonds.net/files/files/How-to-Issue-Green-City-Bonds.pdf
Cheshmehzangi A, Butters C, Xie L, Dawodu A (2021) Green infrastructures for urban sustainability: issues, implications, and solutions for underdeveloped areas. Urban For Urban Green, 59:127028. https://doi.org/10.1016/j.ufug.2021.127028
ChinaBond (n.d.-a) Evaluation report for the Jiangxi Government Green Special Bond [Dataset]. https://www.chinabond.com.cn/dfz/#/
ChinaBond (n.d.-b) Pre-issuance disclosure documents for the 'Guangdong Government Green Special Bond' [Dataset]. https://www.chinabond.com.cn/dfz/#/
City of Vancouver (2018) 3.10% sinking fund debentures series. https://vancouver.ca/files/cov/offering-circular-due-2028.pdf
City of Vancouver (n.d.-a) City of Vancouver sustainability bond framework. https://vancouver.ca/files/cov/vancouver-sustainalytic-sustainability-bond-framework.pdf
City of Vancouver (n.d.-b) Investor relations. https://vancouver.ca/your-government/investor-relations.aspx
CPI. (2023). *Financial Aggregation Blueprints for Urban Climate Infrastructure*. https://www.climatepolicyinitiative.org/wp-content/uploads/2023/06/Annex-2-Municipal-Green-Bonds-Blueprints.pdf
Dembele F, Schwarz R, Horrocks P (2021) Scaling up green, social, sustainability and sustainability-linked bond issuances in developing countries (OECD Development Perspectives 11; OECD Development Perspectives, Vol. 11). https://doi.org/10.1787/8a5c3156-en
European Parliament. Directorate General for Internal Policies of the Union (2022) Green bonds: an assessment of the proposed EU Green Bond Standard and its potential to prevent greenwashing. Publications Office. https://data.europa.eu/doi/10.2861/26821
Flammer C (2021) Corporate green bonds. J Financ Econ 142(2):499–516. https://doi.org/10.1016/j.jfineco.2021.01.010
Flyvbjerg B (2011) Case study. In: The Sage handbook of qualitative research, 4th edn, pp 301–316. https://papers.ssrn.com/sol3/papers.cfm?abstract_id=2278194
Forsgren K (2016) What's next for U.S. Municipal Green Bonds? https://www.spglobal.com/en/research-insights/articles/Whats-Next-For-US-Municipal-Green-Bonds
Ghaemi Asl M, Smutka L, Nasr Isfahani M, Raza SA, Vasa L (2024) Can multifarious types of green bonds be accused of greenwashing with a durative analysis? Insights from a permanent causality vs. temporary causality phenomenon. Environ Dev Sustain. https://doi.org/10.1007/s10668-024-04501-z
Gorelick J, Cara E, Kavoo G (2024) The fallacy of green municipal bonds in developing countries. WORLD 5(4):929–951. https://doi.org/10.3390/world5040047
Herrera H (2024) The proliferation of municipal green bonds in Africa and Latin America: the need for a climate justice approach. Environ Urban 36(1):147–172. https://doi.org/10.1177/09562478241230290
Hong Kong Monetary Authority (2021) The Hong Kong Special Administrative Region Government issues green bonds. https://www.hkma.gov.hk/gb_chi/news-and-media/press-releases/2021/11/20211118-3/
Huan S, Bridget B, Yanjing W, Yuqing L, Rui Z (2020) China's green bond issuance and investment opportunity report. https://www.climatebonds.net/files/reports/cbi_gfo_china_05b.pdf

ICMA (2021) Guidance handbook. https://www.icmagroup.org/assets/documents/sustainable-finance/2021-updates/the-gbp-guidance-handbook-june-2021-140621.pdf
ICMA (2024) Transition finance in the debt capital market. https://www.icmagroup.org/assets/Transition-Finance-in-the-Debt-Capital-Market-paper-ICMA-14022024.pdf
ICMA (n.d.) Green bond principles. International Capital Market Association. https://www.icmagroup.org/assets/documents/Sustainable-finance/2022-updates/Green-Bond-Principles_June-2022-280622.pdf
Kahagalle H, Lindeman-Jones A (2021) Bonding over Green—Benefits of Green Bonds for issuers and investors. https://addisons.com/knowledge/insights/bonding-over-green-benefits-of-green-bonds-for-issuers-and-investors/
Karpf A, Mandel A (2017) Does it pay to be green? SSRN Electron J. https://doi.org/10.2139/ssrn.2923484
Liu W, Liu H (2024) Green and sustainable municipal bonds: new drivers for infrastructure investment. https://www.greenpeace.org.cn/wp-content/uploads/2024/04/Green-Municipal-Bond-Policy-Briefing.pdf
MoF (2024) Japan Climate Transition Bonds [Dataset]. https://www.mof.go.jp/english/policy/jgbs/topics/JapanClimateTransitionBonds/index.html
OECD (2017a) Investing in climate, investing in growth. OECD. https://doi.org/10.1787/9789264273528-en
OECD (2017b) Mobilising bond markets for a low-carbon transition. OECD. https://doi.org/10.1787/9789264272323-en
OECD (2023) Mechanisms to prevent carbon lock-in in transition finance. OECD. https://doi.org/10.1787/d5c49358-en
OECD (2024) Infrastructure for a climate-resilient future. OECD. https://doi.org/10.1787/a74a45b0-en
OECD, The World Bank, & United Nations Environment Programme (2018) Financing climate futures: rethinking infrastructure. OECD. https://doi.org/10.1787/9789264308114-en
Packer F, Ehlers T (2017) Green bond finance and certification. BIS Quarterly Review. https://www.bis.org/publ/qtrpdf/r_qt1709h.htm
People's Bank of China (2021) Catalogue of projects supported by Green Bonds (2021 edition). https://www.gov.cn/zhengce/zhengceku/2021-04/22/content_5601284.htm
Shishlov I, Morel R, Cochran I (2016) Beyond transparency: Unlocking the full potential of green bonds. https://www.cbd.int/financial/greenbonds/i4ce-greenbond2016.pdf
Shrimali G (2021) Transition bond frameworks: goals, issues, and guiding principles. https://sfi.stanford.edu/sites/sfi/files/media/file/transition_bond_frameworks-_goals_issues_and_guiding_principles_0_0_0_0.pdf
Suzuki M, Yoshitaka M, Milovidova A, Cai H, Yamagata Y (2020) Understanding the potentials of green bonds and green certification schemes for the development of future smart cities. In: Urban systems design. Elsevier, pp 393–407. https://doi.org/10.1016/B978-0-12-816055-8.00013-0
The Government of the & Hong Kong Special Administrative Region (2022) Green Bond framework. https://www.hkgb.gov.hk/en/others/documents/GBF_February_2022_ENG.pdf
The Hong Kong Special Administrative Region Government (2022) Green Bond Report 2022. https://www.hkgb.gov.hk/sc/others/documents/Green_Bond_Report_2022.pdf
True T (2023) Green bonds. https://www.financestrategists.com/wealth-management/bonds/green-bonds/
UN-Habitat (2021) Financing sustainable urban development. https://unhabitat.org/sites/default/files/2021/06/FSUD%20report_web.pdf
Wijaya ME, Siagian AP, Mecca BM, Haesra AR (2021) Accelerating renewable energy finance in Indonesia: The potential of municipal green bonds. https://www.climatepolicyinitiative.org/wp-content/uploads/2021/07/The-potential-of-municipal-green-bonds.pdf

World Bank (2024) Green, social, sustainability, and sustainability linked (GSSS) bonds market update. https://thedocs.worldbank.org/en/doc/dacb969cc71f53abde2d2758f1cc13ed-0340012024/original/GSSS-Quarterly-Newsletter-Issue-No-8.pdf

Yun Z (2020) Market practices and development suggestions for green municipal bonds. https://iigf.cufe.edu.cn/info/1012/1343.htm

Zeng G, Wu Y (2019) Development experience and lessons from U.S. Green Municipal Bonds. https://www.cnfin.com/greenfinance-xh08/a/20190212/1796880.shtml?f=arelated

Zerbib OD (2019) The effect of pro-environmental preferences on bond prices: evidence from green bonds. J Bank Financ 98:39–60. https://doi.org/10.1016/j.jbankfin.2018.10.012

Current Status and Barriers for the Diffusion of Distributed Rooftop Photovoltaics in Southeast Asia: Lessons Learned from Germany, Rwanda, and the United States

Zihao Zhu and Masachika Suzuki

1 Introduction

As distributed rooftop photovoltaic (PV) systems continue to gain global prominence in renewable energy strategies, Southeast Asia (SEA) stands out as a region with immense potential due to its abundant solar resources and growing energy demand. This chapter delves into the current status, barriers, and drivers for the adoption of rooftop PV systems in Malaysia, Thailand, and Vietnam while drawing lessons from Germany, Rwanda, and the United States (US) to offer practical recommendations tailored to SEA's unique context.

This study aligns closely with the overarching goals of the book, which focus on addressing urban and regional challenges through sustainable energy solutions. Distributed rooftop PV systems are particularly relevant as they can enhance energy security, mitigate greenhouse gas emissions, and contribute to economic resilience in urban and regional settings. By identifying the factors influencing the adoption of rooftop PV in SEA, this chapter seeks to provide actionable insights that can guide policymakers and stakeholders in their pursuit of renewable energy goals.

The chapter is structured as follows: The first section categorises the stages of rooftop PV development and policy interventions in Malaysia, Thailand, and Vietnam. The second section examines innovative strategies from Germany, Rwanda, and the US, focussing on how their experiences can address the challenges faced by SEA countries. The concluding section reflects on the future prospects of rooftop PV in SEA, offering policy recommendations to accelerate its adoption.

Z. Zhu (✉)
Global Business Development, TMF Group, Tokyo, Japan
e-mail: zihao.zhu@tmf-group.com

M. Suzuki
Graduate School of Global Environmental Studies, Sophia University, Tokyo, Japan

M. Okitasari et al. (eds.), *Equitable and Sustainable Cities*, Advances in 21st Century Human Settlements, https://doi.org/10.1007/978-981-92-2793-8_8

2 Methodology

To provide a structured and comprehensive analysis, this chapter adopts a comparative case study approach. The selected countries—Malaysia, Thailand, and Vietnam—represent diverse stages of rooftop PV development in SEA, each facing unique barriers and opportunities. Germany, Rwanda, and the US were chosen as reference cases due to their innovative policies and successful deployment of rooftop PV systems under varying socioeconomic and geographic conditions.

Data sources for this analysis include government reports, academic literature, policy briefs, and industry publications. These materials were systematically reviewed to identify key factors influencing rooftop PV adoption, such as financial incentives, regulatory frameworks, technological advancements, and public–private partnerships. Quantitative data, such as installed PV capacity and market penetration rates, were used to complement qualitative insights, ensuring a balanced evaluation.

The chapter is organised into two main parts: The first part focuses on the SEA case studies, detailing the current status, barriers, and policy interventions in Malaysia, Thailand, and Vietnam. The second part examines lessons learned from Germany, Rwanda, and the US, emphasising transferable strategies that can address the challenges faced by SEA countries. By integrating these two components, the methodology ensures that the analysis is both context-specific and globally informed, offering actionable recommendations for policymakers and stakeholders.

With conventional fossil energy reserves rapidly dwindling, renewable energy such as solar and wind need to play a more important role in securing our energy sources. PV power generation technology was born in 1954 and was mainly used in aerospace and industrial construction processes in the early days due to its high cost. After entering the twenty-first century, with the continuous improvement of technology and productivity, the cost of PV power generation has been significantly reduced, and rooftop distributed photovoltaics have gradually become the core force of energy transformation.

Distributed rooftop PV is a form of distributed PV, which is placed on the top floor of each household, converting the irradiated energy of light into direct current (DC) electrical energy and then connected to the alternating current (AC) grid through an inverter. Many developed countries have taken the lead in including rooftop PV in their energy development plans, such as the Million Solar Roofs Program in the EU and the "Rooftop Solar Plan" in the US by Bank of America (Riley 2011). Such programmes are not only designed to solve the problem of daily residential electricity consumption but also to change the centralised pattern of passive reception of energy production and consumption to the active participation of consumers in the self-consumption pattern. The term "Million Solar Roofs" is not a new thing. In SEA, which is more vulnerable to the adverse effects of climate change, further upgrading of renewable energy sources is also widely recognised as a viable solution (Sreenath et al. 2022). Its geography and natural resources, in turn, dictate the importance of photovoltaics—as reference cases already in place—to facilitate the energy transition in SEA. Like many other countries in the world, Southeast Asian economies unveiled their latest climate targets during the Glasgow Climate Conference 2021. With an aggressive vision, Malaysia, Thailand, and Vietnam have chosen distributed rooftop PV as one of the means to achieve their climate commitments, pledging to achieve net-zero emissions between 2050 and 2065.

The next sections analyse the current status and barriers to distributed rooftop PV development in Malaysia, Thailand, and Vietnam, and analyse potential drivers in Germany, Rwanda, and the US, respectively.

The primary objective of this chapter is to provide a comprehensive analysis of the factors influencing the adoption of distributed rooftop PV systems in SEA. By focussing on these three countries, the study aims to explore shared challenges and unique opportunities within the region, such as addressing high upfront costs, limited financing options, and evolving policy frameworks. Through this analysis, the chapter seeks to identify actionable strategies that can support the region's energy transition and contribute to sustainable urban and regional development.

The selection of Malaysia, Thailand, and Vietnam as case studies is motivated by their contrasting levels of PV market maturity and diverse policy approaches. Vietnam represents a rapid growth trajectory driven by favourable Feed-in Tariffs (FiTs), while Thailand showcases gradual policy evolution through mechanisms like Net Energy Metering (NEM). Malaysia, on the other hand, demonstrates a shift from FiTs to a more integrated approach involving fiscal incentives and technology-driven innovations like Peer-to-Peer (P2P) energy trading. These cases provide a rich basis for understanding the dynamics of rooftop PV adoption in SEA.

The inclusion of Germany, Rwanda, and the US as reference cases serves to address research gaps and highlight transferable lessons. Germany's subsidy-driven model offers insights into the design of effective financial incentives, particularly in mature markets. The US exemplifies the role of market mechanisms and mandatory policy goals in driving PV adoption, while Rwanda presents a unique perspective on leveraging innovative financing solutions to overcome resource constraints. By contrasting these contexts, the study not only underscores the adaptability of global best practices but also identifies specific areas where SEA countries can tailor these strategies to their local needs.

3 Stages of Distributed Rooftop PV and Policy Intervention in Southeast Asian Countries

3.1 Vietnam

Vietnam is a successful example of distributed rooftop PV development in Southeast Asia. While the number of solar power installations in Vietnam was almost negligible in 2018, solar power already accounts for a quarter of the total installed capacity of Vietnam's electricity system by the middle of 2021. Distributed rooftop PV has also seen explosive growth. According to the Institute of Energy Economics and Finance

(IEEFA), cumulative rooftop PV installations in Vietnam had been 378 MW by the end of 2019 and 9296 MW by the end of 2020, with more than 101,000 rooftop solar installations in households, offices, and factories in 2022 alone (Thu 2021).

3.1.1 Driving Forces

The direct driver of this solar PV installation boom is the implementation of the Feed-in-Tariff (FIT), which has already expired by the end of 2021 and other complementary policies (Do et al. 2020). FIT allows solar programme investors to sell surplus electricity at a price level above the market rate, which is an attractive investment incentive. Among other things, rooftop solar installations have the highest price, higher than floating solar plants and ground-mounted PV plants. In addition, the implementation of the FIT policy is reinforced by complementary policies that allow solar project developers to raise funds from all sources, including foreign funds, and import duty exemptions for imported equipment (Do et al. 2020). Another interesting point is that the public's willingness to protect the environment and reduce pollution has further contributed to the rapid promotion of the FIT policy (Do et al. 2020). Thu Vu, an analyst for energy finance at the Institute for Energy Economics and Financial Analysis, acknowledges the geopolitical benefit of being close to China as one of the driving forces behind solar growth. About 99% of the solar panels now installed in Vietnam are manufactured in China, and because of their economies of scale and steep learning curve, Chinese manufacturers are able to provide a wide range of solar-related products at low prices (Linh 2021).

3.1.2 Barriers

For rooftop solar projects in Vietnam, one of the bigger hurdles at the moment remains the cost. The cost of rooftop solar project installation is still a large burden compared to Vietnam's GDP per capita (Lee et al. 2019). The lack of trusted rooftop solar system providers and complicated administrative procedures are also barriers to rooftop solar development (Do et al. 2020). After the rooftop solar installation boom, the lack of infrastructure to take on peak power supply—limited transmission grid capacity—and the uncertainty of a subsequent pricing mechanism after the cessation of subsidies are the main causes of wasted capacity at the moment. The planning of transmission lines has been significantly slower than the construction cycle of solar projects, which has resulted in many rooftop solar projects not being able to sell the electricity produced to recoup their investment (Le 2022).

3.2 Thailand

Thailand's favourable tropical location provides a large amount of light radiation, with an average annual radiation output of 1350–1400 kWh/m2, making it an excellent base for solar energy development. According to Thailand's National Power Development Plan (PDP), Thailand's installed PV capacity is expected to reach 15.6 GW by 2037, including 12.7 GW from rooftop PV on homes, and a recent report by The Boston Consulting Group identified distributed rooftop solar as an important tool for realising Thailand's renewable energy transition ambitions (Royal Thai Embassy 2021).

3.2.1 Driving Forces

As one of the first SEA countries to introduce the FIT concept, the Thai government first promoted the development of large-scale PV systems with high-rate subsidies as the main strategy (Chaianong et al. 2019). In its power development plan, the Thai government enabled NEM, a self-consumption mechanism, to stimulate the deployment of household rooftop PV (Emiliano 2019). The government announced the suspension of the FIT policy in 2018 and then announced their latest 25-year FIT plan until 2037 in October 2022 (Santos 2022). The new plan introduced a 25-year FIT of THB 21679 (USD 0.057)/kWh for solar and a 25-year FIT of THB 28331/kWh for solar plus storage. The fact that the net metering programme for rooftop solar PV systems for homes (up to 10 kW) was also launched in May 2019 is noteworthy (Emiliano 2019).

On the fiscal side, the Thai government has established the Energy Service Company Revolving Fund to provide financial support in terms of interest rates, instalment payments, and credit limits. Using this fund, financial aid was provided from 2015 to 2018 with an interest rate of 3.5%, a five-year instalment plan, and a credit line (within 50 million THB) (Energy for Environmental Foundation 2022). It also provides tax incentives such as tax exemptions on imported equipment and income tax exemptions for participating companies in rooftop solar for a specified period (Sreenath et al. 2022). On the non-tax side, the Thai government allows foreign funds to have 100% equity ownership of solar projects. They also simplified the foreign exchange remittance system and opened up permission for foreign investors to purchase land. These initiatives have undoubtedly attracted large amounts of foreign capital for investment (Thailand International Trade Administration 2021).

Notably, Thailand launched a pilot Peer-to-Peer (P2P) renewable energy trading in August 2018 and has been operating successfully until now. This new decentralised micro-power trading model could be a potential catalyst for a shift to a new paradigm in the energy mix (Pinyo et al. 2021). The P2P energy trading model encourages the creation of resilient micro-grid deployments through a market-based trading mechanism that operates in tandem with government subsidy policies.

3.2.2 Barriers

Policy changes that are fraught with uncertainty as well as discontinuous types remain a major impediment to the development of rooftop solar PV (Chatthaworn et al. 2018). Changes in the number of on-grid tariffs and financial incentives can lead to difficulties in valuing projects, and this severely affects investor confidence. In addition, similar to the situation in other developing countries, the upfront cost of rooftop solar is likely to be an unaffordable expense for potential holders. It is worth noting that some scholars have mentioned that over-reliance on imported technology and equipment will inhibit the growth and subsequent development of domestic industry and knowledge base (Chaianong and Pharino 2015). This is because it will lead to a long-term lack of sufficiently skilled labour and production capacity for solar PV, which is essential for a sustainable solar energy system.

3.3 Malaysia

Malaysia, located in the centre of SEA, is an excellent country for developing solar energy, with excellent geographical conditions and a light intensity of 1400–1900 kW h/m^2 (Lau et al. 2022). The Malaysian government has been emphasising the importance of increasing the development of rooftop PV power generation in recent years. According to Sustainable Energy Development Authority Malaysia, if all the houses in Peninsular Malaysia were topped with solar panels, they would be able to generate enough electricity to meet the current demand. With over 4.12 million buildings in Malaysia, there is great potential for rooftop solar development. If all of them are installed with solar PV systems, it is expected to reach a capacity of 34,194 MW (Sustainable Energy Development Authority 2019).

3.3.1 Driving Forces

In the early stages of the PV industry development, the Malaysian government decided to use the FIT policy as the main incentive to advance PV deployment and development. However, due to the pressure on the use of the subsidy, the Malaysian government ended this policy in late 2017 and switched to the Net Energy Metering (NEM) programme as a new incentive. The core of the NEM policy is the desire to reduce the fiscal burden by implementing self-generation of electricity (Sustainable Energy Development Authority 2022). In the NEM 2.0 programme, a "one-to-one" compensation mechanism was introduced to make each kilowatt hour injected into the grid offset against the one kilowatt hour of electricity taken from the grid. In addition, Malaysia is also participating in a global pilot project for P2P energy trading in 2019 (Husain et al. 2021). In terms of fiscal incentives, participants in the PV industry in Malaysia are also eligible for Green Investment Tax Allowance (GITA) and Green Income Tax Exemptions (GITE) (Malaymail 2020). Companies that provide solar leasing services are entitled to a 70% income tax exemption for up to 10 years, according to Budget 2020. A 100% investment tax allowance and income tax exemption could be available for some RE programmes. Additionally, Malaysian-owned companies can get financial support for their RE projects through the Green Technology Financing Scheme (GTFS).

3.3.2 Barriers

One of the key levers for the development of rooftop PV is infrastructure readiness. This is especially true for Malaysia, which is already involved in P2P. Combining P2P has the potential to reduce electricity congestion, providing a centralised residential management control system. In addition, the cost of electricity storage is higher than that of electricity in Malaysia. Although efforts have been made in Malaysia through a complementary energy storage system (ESS) (Heng et al. 2022), the affordability of ESS in the future will be a potential impediment to promoting the development of rooftop PV. A sustainable rooftop PV market cannot be created without implementing a uniform and standardised policy in the technology application area. Lack of healthy competition in distribution channels to provide satisfactory service to PV system adopters may also be a problem that Malaysia will face in the future (Lau et al. 2022).

3.4 Summary

Table 1 summarises the key support mechanisms for rooftop PV development in Vietnam, Thailand, and Malaysia, including Feed-in Tariffs (FiTs), Net Energy Metering (NEM), Peer-to-Peer (P2P) trading, Renewable Portfolio Standards (RPS), and various fiscal incentives. While these mechanisms share common goals, their implementation and effectiveness vary significantly across the three countries.

Vietnam's reliance on FiTs has been instrumental in driving the rapid growth of rooftop PV installations. The FiT mechanism provided attractive investment incentives by allowing investors to sell surplus electricity at above-market rates. However, the expiration of FiTs in 2021 created uncertainty, leaving the market in a transitional phase. Fiscal incentives such as import duty exemptions also played a supportive role but were insufficient to address systemic challenges like grid infrastructure bottlenecks and administrative barriers.

Thailand's energy policies have evolved from high-rate FiTs to the introduction of NEM schemes. The NEM policy incentivises self-consumption, reducing reliance on government subsidies. Complementary fiscal policies, such as the Energy Service Company (ESCO) Revolving Fund and tax exemptions for imported equipment, have attracted both domestic and foreign investment. Thailand's pilot P2P energy trading projects further exemplify innovative approaches to decentralised energy management, although these remain in nascent stages.

Table 1 Rooftop PV-related support mechanism

Country	Rooftop PV-related support mechanism (As of October 2022)				
	FiT	NEM	P2P	RPS	Fiscal incentives
Vietnam	Expired	\	\	\	\
Thailand	For Year 2022–2037: 25-year FIT of THB 21679 (USD 0.057)/kWh for solar 25-year FIT of THB 28331/kWh for solar plus storage	For residential rooftop solar PV systems (up to 10 kW)	Pilot phase: Project covered 635 kW capacity, which was implemented across the T77 precinct over two years duration	\	Energy service company revolving fund Tax exemptions on imported equipment Income tax exemptions
Malaysia	Expired	NEM 3.0 (will be in effect from 2021 to 2023): NEM Rakyat Programme 100 MW quota allocation NEM GoMEn Programme 100 MW quota allocation NOVA Programme 600 MW quota allocation	Pilot phase: Alpha run and Beta run	\	Green Investment Tax Allowance (GITA) Green Income Tax Exemptions (GITE) Green Technology Financing Scheme (GTFS)

Malaysia's transition from FiTs to the NEM programme highlights its focus on reducing fiscal burdens while promoting self-generation. Programmes like NEM 3.0 include features such as quota allocations and one-to-one compensation mechanisms, ensuring fair returns for participants. Fiscal incentives, including the Green Investment Tax Allowance (GITA) and Green Technology Financing Scheme (GTFS), support industry players and residential adopters alike. Malaysia's efforts in piloting P2P trading models also reflect its commitment to leveraging technology for energy democratisation.

A closer examination of Table 1 reveals notable trends and gaps. While FiTs have proven effective in jumpstarting PV adoption, their abrupt discontinuation can destabilise markets, as seen in Vietnam. NEM policies in Malaysia and Thailand offer more sustainable pathways but require robust grid infrastructure and regulatory frameworks to ensure long-term viability. Fiscal incentives across all three countries underscore the importance of financial support, yet their impact is limited without parallel investments in technical capacity and market development.

The integration of innovative mechanisms like P2P trading in Malaysia and Thailand indicates a shift towards more decentralised and consumer-centric energy systems. However, these models require careful scaling and regulatory oversight to avoid market fragmentation. Policymakers in SEA can draw lessons from the nuanced implementation of these mechanisms to create tailored strategies that address both immediate and structural challenges in rooftop PV deployment.

Compared to the developed countries, although the policy environment in Southeast Asia countries provides many favourable conditions for the diffusion of distributed solar energy systems, further promotion of the systems faces several common barriers as follows:

Soft costs of installing PV are too high, resulting in relatively high costs for distributed rooftop PV. The soft costs include financing costs, complex administrative procedures, information acquisition costs, installation, and operations and management (O&M) costs. Local authorities and grid companies may also need to negotiate to obtain permits before installation.

There are limited financing options. Banks perceive solar assets as having high revenue risk and long payback periods. The cost of rooftop solar remains high for residents in emerging markets, and the lack of access to low-cost capital support from a mature consumer lending market also hinders solar financing.

There is inadequate planning to carry electricity through the grid system. The grid planning is much slower than the construction of new solar projects, resulting in outdated grid plans that are unable to carry the power generated by new PV projects, which affects the ROI of the projects because their power cannot be fed into the grid.

4 Lessons from Germany, the United States, and Rwanda

From the analysis, we recognise that high financing costs and limited financing channels are the inescapable obstacles to the development of distributed rooftop PV in SEA countries. Therefore, the reasonable use of market mechanisms to obtain low-cost financing becomes the key to achieving further widespread promotion. Due to the basic characteristics of rooftop PV projects with the large initial investment, long payback period, and long-term stable cash flow, some countries have attempted to securitise PV assets. PV securitisation refers to the act of converting PV plant assets into asset-backed securities that can be freely traded in financial markets, making them liquid (Alafita and Pearce 2014). This section addresses the potentials and insights of rooftop PV securitisation using a successful project in Rwanda as an example. Additionally, presenting lessons from Germany and the US adds valuable insights into effective policy frameworks and financial models, which could help overcome similar challenges in SEA, where PV adoption faces financing barriers.

4.1 Germany

Germany is one of the leading countries in renewable energy development with 97% of its renewable energy generation being integrated into the distribution grid and utilised in a distributed manner. In Germany, 30% of households have already installed rooftop solar panels, distributed PV power accounted for about 8% of the country's annual electricity consumption, and rooftop PV power systems accounted for about 85% of the total installed PV capacity in 2017 (Bayer et al. 2018). The high electricity prices and the strength of policy subsidies are the most important factors driving the development of distributed rooftop PV in the country (BOC Securities 2022). And the extremely fine-grained route of subsidy rebate rate adjustments made the installed capacity curve flatter, giving the finance department ample time to adjust and plan the next policy step. The initial decline in PV subsidies was on an annual basis, which was changed from annual to quarterly starting in June 2010 and then to a monthly basis starting in April 2012 (Wirth 2021).

The relevant policy support in Germany is divided into a PV fixed Feed-in tariff (FIT), Feed-in Premium (FIP), and inclusive financing approaches. The Feed-in Tariff for rooftop PV is substantially higher than the market tariff, and some self-generating plant operators are eligible for surcharge reductions. FIP will fluctuate with the wholesale price in the electricity market, with the grid operator paying the difference between the market value and the reference value as a subsidy, aiming to reduce costs through market competition and thus reducing the burden on the public. In terms of financing channels, German distributed rooftop PV power plants are financed by German policy banks, local commercial banks, and power plant investors, which directly provide lower-cost loans. For smaller-scale rooftop PV power plants, no equity collateral is required to apply for a loan. FIT revenues and PV project assets are used as collateral directly after the bank's credit assessment of the owner (te Heesen et al. 2019).

In addition, several new policy overlays have further boosted the development of distributed household rooftop PV. EEG 2021 (Germany's Landmark Renewable Energy Act 2021) proposes that self-consumption PV with an installed capacity of up to $300 million will be exempt from the EEG tax (renewable energy tax) (IEA 2021). The cap on PV generation will be lifted in 2020, and PV projects whose construction is blocked by COVID will be allowed to delay completion. The Berlin Solar Law, adopted in 2021, requires all new buildings in Berlin to install PV systems from 2023 and to cover at least 30% of the roof area (Spiegel 2021). Significant simplification of the reporting process and administrative procedures for the construction of rooftop PV projects has also become a major stimulus for the development of rooftop PV.

For the mismatch between lagging grid planning and the high incremental solar grid connection, Germany has used a reasonable increase in the load factor of the power-rich selected grid as a transitional solution (Rechsteiner 2021). In addition, the German transmission grid operators (TSOs) typically propose grid planning scenarios ten years in advance, forecast future electricity production and consumption through scenario frameworks, and revise development plans with public participation. The German federal government has designed six time points, each corresponding to a different phase of grid planning. Time targets are agreed upon between the German government and the TSOs and are required to be realistic and not based on "best case" assumptions. The progress of the projects will be monitored every three months, and measures will be taken for potential delays (BOC Securities 2022).

4.2 United States

Despite being an early adopter of the PV industry, the widespread deployment of residential rooftop solar has only been completed in the last decade. Since 2005, installed residential solar capacity has grown more than seventyfold. As of 2015, residential rooftop PV systems accounted for nearly one-third of the total installed capacity (O'Sullivan and Warren 2016). Over the past decade, the dramatic reduction in hardware costs for PV and incentives represented by the investment tax credit (ITC)—called Solar Investment Tax Credit—have been the core drivers of the growth in installed PV capacity in the US (Adams 2020).

ITC is a fiscal incentive offered by the federal government to encourage taxpayers to invest in renewable energy generation equipment. For qualifying solar installations, the government will reduce the tax liability for investors by 30% of the cost of the equipment. The Inflation Reduction Act of 2022 extends the 30% ITC level through 2032 to continue incentivising clean energy deployment (Internal Revenue Service 2022).

The growing risk of power outages has also become another stimulus for the rapid growth in the installation of rooftop PV systems. The US energy and electricity market has more of a commodity attribute in a free market than a public utility attribute. The US grid was built at its peak before the 1980s and will be ageing by 2021, with the reliability of power supply relying on stand-alone residential energy storage systems (Patel et al. 2021). After the time of massive power outages in Texas, affected by the snowstorm in February 2021, there was a spike in the search for PV modules locally and nationwide, indicating that more residents are considering the installation of household rooftop PV storage systems to cope with possible large-scale power outages (Lewis 2022).

While US renewable energy-related subsidies are subject to uncertainty due to regime change, a review of the history of the US renewable energy industry shows that incentives have been more resilient. With policy goals such as "carbon-free power," even a temporary suspension of ITC subsidies is likely to be reinstated in the future (BOC Securities 2022).

Another important factor is the Renewable Portfolio Standard (RPS). The goal of RPS is set individually by each state based on local conditions, effectively guaranteeing the development of renewable energy goals in each region. As of June 2022, 24 states that are members of the US Climate Alliance have announced that they will achieve 73% renewable energy generation by 2040. According to statistics, 31 states and two special districts in the US have established clear renewable energy development goals (National Conference of State Legislatures 2022). The states' energy targets are constantly updated upward, showing the regions' firm determination to achieve low-carbon goals and support the high growth of the PV industry from a policy perspective.

4.3 Rwanda

Rwanda's PV securitisation project facilitated by the Global Innovation Lab for Climate Finance is located in East Africa with abundant sunshine (over 24 million hours/m^2). Only 34% of Rwanda's nearly 13 million people have access to the national grid, so one of the main objectives of this project is to boost the base of off-grid PV plants (Global Innovation Lab for Climate Finance 2019). Funding the market for distributed rooftop PV systems by forming asset-backed securities from the future earnings of multiple solar developers expands the pool of solar developers and accelerates advanced flows. Figure 1 illustrates the mechanisms of solar securitisation.

Through the involvement of this innovative financial instrument, several significant obstacles affecting the existing development of rooftop PV could be overcome as follows:

1. In the previous financing approach, banks often required collateral worth 1.25 times the value of the loan to be granted (Global Innovation Lab for Climate Finance 2019), and solar companies were often undercapitalised. In this market-based financing approach, individual and institutional investors' capital is mitigated by securitisation to alleviate the high demand for collateral.
2. Previously, the loan would need to be repaid within 2–5 years (Global Innovation Lab for Climate Finance 2019), and the outstanding balances would be recorded as assets on the solar company's balance sheet. Securitisation, on the other hand, is a form of off-balance sheet financing that provides off-balance sheet financing

Fig. 1 Mechanisms of solar securitisation. *Source* Global Innovation Lab for Climate Finance

by securitising the future cash flows of PV assets, increasing the leverage and risk tolerance of the solar company.

3. Funding for solar companies no longer depends solely on local credit lines and total funding, and government fiscal spending has been further eased. PV securitisation has allowed solar companies to make greater use of liquid capital markets to access lower-cost external market funding.

The uncertainty, non-standardisation, and cross-liquidity of solar projects have led to high interest rates on solar loans. Through asset securitisation, where risks are packaged and sold separately by grade to investors with different needs, solar companies often have cost-competitive interest rates for subsequent issues once they have paid the initial set-up costs of asset-backed securities (Global Innovation Lab for Climate Finance 2019).

5 Reflection and Prospects for Rooftop Distributed PV in Southeast Asia

As seen from the German experience, long-term planning and sufficiently detailed rebate routes can effectively slow down the government's fiscal spending curve and keep rooftop solar installations in an orderly and stable state. In addition, compared to a direct FIT cut-off, an FIP with a floating price mechanism can effectively reduce costs through market competition when the market matures. For countries like Vietnam that have discontinued the FIT mechanism and are considering a new PPA model (Thompson et al. 2022), the participation of FIP may make the policy adjustment period smoother. The main factor in the success of distributed solar development in the US is the implementation of the ITC policy, which highlights the role of financial incentives in the promotion of distributed solar. In addition, mandatory policy goals, as represented by the RPS, can be introduced gradually as the market matures. For instance, it is stated that the RPS mechanism will be gradually deployed to large power-producing businesses under Vietnam's Renewable Energy Development Strategy (2016–2030). Companies that produce more than 1000 MW must source 10% of their electricity from renewable sources by 2030 (IEA 2016).

At present, the funds involved in the PV industry in SEA countries are mainly public funds, such as finance, and the potential of social capital has not been fully released. SEA countries may consider trying to replicate Rwanda's PV securitisation model, using developmental financial institutions to guide and credit-enhance the role of future earnings from solar projects to form asset pools, leveraging commercial investors' funds into it through off-balance sheet financing. However, it should be noted that although the use of financial engineering financing can effectively avoid greater pressure on the financial side, comprehensive laws and regulations and strong supervision and management are often needed to guide the healthy development of asset securitisation. In addition, the equipment of PV projects may generate quality or O&M problems after a period of operation, leading to a continuous decline in power generation efficiency and output, resulting in lower-than-expected project cash flow. Therefore, SEA countries should consider how to construct sufficiently clear legal regulations before introducing PV securitisation and implement comprehensive measures in terms of mechanism innovation and mechanism progress. At the same time, they can continue to promote technology and equipment innovation, reduce the cost of PV projects, and reduce project costs and expenses. Another important point is the need to improve the management level of PV project pre-development and post-operation and maintenance to ensure the most stable power generation efficiency and capacity as possible.

Declaration of Competing Interests The research described in this article is supported by the MEXT/JSPS KAKENHI Grant Number JP20K01917. The original version of this article was presented at NERPS Conference 2023: Bridging Peace and Sustainability amidst Global Transformations (online) in 2023 in Thailand.

References

Adams R (2020) How the solar tax credit makes renewable energy affordable. https://na.panasonic.com/us/green-living/how-solar-tax-credit-makes-renewable-energy-affordable

Alafita T, Pearce JM (2014) Securitization of residential solar photovoltaic assets: costs, risks and uncertainty. Energy Policy 67:488–498. https://doi.org/10.1016/j.enpol.2013.12.045

Bayer B, Matschoss P, Thomas H, Marian A (2018) The German experience with integrating photovoltaic systems into the low-voltage grids. Renew Energy 119:129–141. https://doi.org/10.1016/j.renene.2017.11.045

BOC Securities (2022) In-depth analysis of overseas PV market "海外光伏市场深度解析." https://pdf.dfcfw.com/pdf/H3_AP202206071570557934_1.pdf?1654607798000.pdf

Chaianong A, Bangviwat A, Menke C (2019) Forecast of rooftop PV adoption in Thailand. Sci Technol Asia, 13–41

Chaianong A, Pharino C (2015) Outlook and challenges for promoting solar photovoltaic rooftops in Thailand. Renew Sust Energ Rev 48:356–372. https://doi.org/10.1016/j.rser.2015.04.042

Chatthaworn R, Angaphiwatchawal P, Chaitusaney S (2018) Solar PV policy, barriers and proposed solution for technical barriers in Thailand. https://www.researchgate.net/publication/330637083_Solar_PV_Policy_Barriers_and_Proposed_Solution_for_Technical_Barriers_in_Thailand

Do TN, Burke PJ, Baldwin KGH, Nguyen CT (2020) Underlying drivers and barriers for solar photovoltaics diffusion: the case of Vietnam. Energy Policy 144:111561. https://doi.org/10.1016/j.enpol.2020.111561

Emiliano B (2019) Thailand launches net metering scheme for residential PV. PV Magazine International. https://www.pv-magazine.com/2019/05/24/thailand-launches-net-metering-scheme-for-residential-pv/

Energy for Environmental Foundation (2022) ESCO revolving fund. http://www.efe.or.th/escofund.php?task=&sessid=&lang=en

Global Innovation Lab for Climate Finance (2019) Solar securitization for Rwanda. The Global Innovation Lab for Climate Finance. https://www.climatefinancelab.org/project/solar-securitization/

Heng YB, Ramachandaramurthy VK, Verayiah R, Walker SL (2022) Developing Peer-to-Peer (P2P) energy trading model for Malaysia: a review and proposed implementation. IEEE Access 10:33183–33199. https://doi.org/10.1109/ACCESS.2022.3161577

Husain AAF, Ahmad Phesal MH, Kadir MZAA, Ungku Amirulddin UA, Junaidi AHJ (2021) A decade of transitioning Malaysia toward a high-solar PV energy penetration nation. Sustainability 13(17):Article 17. https://doi.org/10.3390/su13179959

IEA (2016) Vietnam renewable energy development strategy 2016–2030 with outlook until 2050 (REDS)—policies. IEA. https://www.iea.org/policies/6095-vietnam-renewable-energy-development-strategy-2016-2030-with-outlook-until-2050-reds

IEA (2021) Germany's Renewables Energy Act 2021—Policies. IEA. https://www.iea.org/policies/12392-germanys-renewables-energy-act

Internal Revenue Service (2022) Inflation Reduction Act of 2022 | Internal Revenue Service. https://www.irs.gov/inflation-reduction-act-of-2022

Lau L-S, Choong Y-O, Ching S-L, Wei C-Y, Senadjki A, Choong C-K, Seow A-N (2022) Expert insights on Malaysia's residential solar-energy policies: shortcomings and recommendations. Clean Energy 6(4):619–631. https://doi.org/10.1093/ce/zkac043

Le L (2022) After renewables frenzy, Vietnam's solar energy goes to waste. Earth Journalism Network https://earthjournalism.net/stories/after-renewables-frenzy-vietnams-solar-energy-goes-to-waste

Lee N, Cardoso de Oliveira R, Roberts B, Katz J, Brown T, Flores-Espino F (2019) Exploring renewable energy opportunities in select Southeast Asian countries: a geospatial analysis of the levelized cost of energy of utility-scale wind and solar photovoltaics (NREL/TP-7A40-71814). National Renewable Energy Lab (NREL), Golden, CO (United States). https://doi.org/10.2172/1527336

Lewis M (2022) Here's how rooftop solar could have prevented 2021's Texas big freeze power outage. Electrek. https://electrek.co/2022/02/11/heres-how-rooftop-solar-could-have-prevented-2021s-texas-big-freeze-power-outage/

Linh P (2021) China key to Vietnam's solar success. China Dialogue. https://chinadialogue.net/en/energy/china-key-to-vietnams-solar-success/

O'Sullivan FM, Warren CH (2016) Solar securitization: An innovation in renewable energy finance. Main. https://energy.mit.edu/publication/solar-securitization-innovation-renewable-energy-finance/

Malaymail (2020) Minister: Malaysia set to maximise green industry, renewable energy potential in 2020 | Malay Mail. https://www.malaymail.com/news/malaysia/2020/01/03/minister-malaysia-set-to-maximise-green-industry-renewable-energy-potential/1824394

National Conference of State Legislatures (2022) State renewable portfolio standards and goals. https://www.ncsl.org/research/energy/renewable-portfolio-standards.aspx

Patel S, Ceferino L, Liu C, Kiremidjian A, Rajagopal R (2021) The disaster resilience value of shared rooftop solar systems in residential communities. Earthquake Spectra 37(4):2638–2661. https://doi.org/10.1177/87552930211020020

Pinyo A, Bangviwat A, Menke C, Monti A (2021) Decentralized P2P electricity trading model for Thailand. Sensors (Basel, Switzerland) 21(21):7413. https://doi.org/10.3390/s21217413

Rechsteiner R (2021) German energy transition (Energiewende) and what politicians can learn for environmental and climate policy. Clean Techn Environ Policy 23(2):305–342. https://doi.org/10.1007/s10098-020-01939-3

Riley C (2011) Bank of America rooftop solar project is world's biggest. CNNMoney. https://cnnmoney.com/2011/06/22/news/companies/bank_of_america_solar/index.htm

Royal Thai Embassy, Washington DC (2021) Boston consulting: Rooftop solar key for Thailand. https://thaiembdc.org/2021/10/18/boston-consulting-rooftop-solar-key-for-thailand/

Santos B (2022) Thailand introduces FIT scheme for solar, storage. PV Magazine International. https://www.pv-magazine.com/2022/10/31/thailand-introduces-fit-scheme-for-solar-storage/

Spiegel (2021) City of Berlin lawmakers pass rooftop solar law, approve buyback of city grid. Clean Energy Wired. https://www.cleanenergywire.org/news/city-berlin-lawmakers-pass-rooftop-solar-law-approve-buyback-city-grid

Sreenath S, Azmi AM, Dahlan NY, Sudhakar K (2022) A decade of solar PV deployment in ASEAN: policy landscape and recommendations. Energy Rep 8:460–469. https://doi.org/10.1016/j.egyr.2022.05.219

Sustainable Energy Development Authority, Malaysia (2019) Malaysia can generate more electricity if all roofs use solar panels, says Yeo. https://www.seda.gov.my/2019/05/malaysia-can-generate-more-electricity-if-all-roofs-use-solar-panels-says-yeo/

Sustainable Energy Development Authority, Malaysia (2022) NEM 2.0—Renewable Energy Malaysia. https://www.seda.gov.my/reportal/nem2/

te Heesen H, Herbort V, Rumpler M (2019) Performance of roof-top PV systems in Germany from 2012 to 2018. Sol Energy 194:128–135. https://doi.org/10.1016/j.solener.2019.10.019

Thailand International Trade Administration (2021) Energy resource guide—Thailand—Renewable energy. https://www.trade.gov/energy-resource-guide-thailand-renewable-energy

Thompson B, Harrison D, Chow MB (2022) Vietnam's direct PPA pilot scheme | Energy market update—June 2022 | Perspectives & events | Mayer Brown. https://www.mayerbrown.com/en/perspectives-events/publications/2022/06/vietnams-direct-ppa-pilot-scheme-energy-market-update-june-2022

Thu V (2021) IEEFA: Vietnam's extraordinary rooftop solar success deals another blow to the remaining coal pipeline. https://ieefa.org/resources/ieefa-vietnams-extraordinary-rooftop-solar-success-deals-another-blow-remaining-coal

Wirth H (2021) Recent facts about photovoltaics in Germany—Fraunhofer ISE. Fraunhofer Institute for Solar Energy Systems ISE. https://www.ise.fraunhofer.de/en/publications/studies/recent-facts-about-pv-in-germany.html

How Innovative Financial Technologies Can Foster Equity and Sustainability in Cities and Regions

Mohamad Nadim Adi and Mais M. Aljunaidy

1 Introduction

Financial technology, commonly referred to as fintech, is radically changing how financial services are accessed, delivered, and managed. Through a blend of digital innovations ranging from mobile payments to artificial intelligence (AI), the impact of fintech stretches beyond the financial sector. By addressing systemic inefficiencies and fostering sustainability, fintech's transformative potential extends into urban and regional planning, making it a powerful tool to tackle pressing challenges such as resource management, environmental stewardship, and socio-economic disparities. Fintech has the potential to fundamentally reshape socio-economic systems by making financial services more accessible, inclusive, and efficient. One of the most pressing challenges within the financial sector is financial exclusion, particularly in developing countries. According to the World Bank, about 1.7 billion adults globally are unbanked, limiting their economic mobility and exacerbating socio-economic inequalities (Demirguc-Kunt et al. 2018). Without access to crucial financial services such as credit, savings accounts, and investments, individuals in these regions face considerable barriers to improving their financial security and climbing the economic ladder. Traditional banking institutions are often unable to cater to low-income populations due to high operational costs and regulatory restrictions. Serving rural or remote areas has proven unviable for many traditional banks. However, fintech offers a scalable and cost-effective alternative by bypassing the need for brick-and-mortar branches. Beyond financial access, fintech's innovative

M. N. Adi (✉)
Interior Design, School of Family and Consumer Sciences, Texas State University, San Marcos, TX, USA
e-mail: nadim.adi@txstate.edu

M. M. Aljunaidy
Department of Psychology, College of Liberal Arts, Texas State University, San Marcos, TX, USA

M. Okitasari et al. (eds.), *Equitable and Sustainable Cities*, Advances in 21st Century Human Settlements, https://doi.org/10.1007/978-981-92-2793-8_9

delivery models also empower local governments and non-governmental organisations (NGOs) to implement targeted interventions, such as distributing financial aid or subsidies directly to marginalised populations, thereby enhancing the efficiency of urban and regional governance. As the digital revolution advances, fintech's potential to drive equitable growth and sustainability becomes more apparent. Fintech provides easy and fast access to automated financial processes. For example, mobile applications such as Robinhood, Mint, Coinbase, and Venmo have made it easier than ever to transfer funds, using just a smartphone. Mobile money platforms, like Kenya's M-Pesa, have proven to be game-changers in advancing financial inclusion. M-Pesa allows users to send, receive, deposit, and withdraw money using mobile phones, bringing banking services to millions of people who previously had no access to them (Suri and Jack 2016). Despite the platform's success, it still relies on intermediaries such as mobile network operators, which may restrict scalability. Moreover, while M-Pesa has had substantial impacts on individual financial security, its success highlights the broader potential of fintech to strengthen regional economies by integrating informal sectors into formal financial systems and stimulating entrepreneurial growth. For example, M-Pesa has enabled farmers in rural Kenya to access micro-loans via partnerships with agricultural cooperatives, enhancing productivity and connecting rural producers with urban markets through digital payments.

A recent and potentially more scalable solution lies in blockchain and digital currencies, which offer a decentralised, peer-to-peer alternative to traditional banking systems. By removing intermediaries, blockchain can significantly increase transaction speed, reduce transaction costs, and enhance the security of financial transfers. This is especially beneficial for migrant and low-income workers sending remittances, as blockchain technology can bypass costly remittance services, ensuring that a larger portion of their earnings reaches their families. In urban areas, blockchain is being used to create transparent municipal financial systems, enabling cities to track public spending and ensure accountability. For instance, Dubai's Blockchain Strategy aims to digitise and secure government records, including land registries and building permits, reducing corruption and improving urban governance efficiency (Digital Dubai 2020). However, the complexity of blockchain technology and the digital literacy required to manage digital wallets can present barriers to its widespread adoption, particularly in developing regions (Foley et al. 2019).

Sustainability is another key area where fintech can make a significant impact. Although cryptocurrency mining has been widely criticised for its energy-intensive processes, it has sparked innovative solutions. One such solution involves using surplus renewable energy to power mining operations. In regions with an abundance of renewable energy resources, such as Iceland and Norway, cryptocurrency mining is powered by geothermal and hydroelectric energy, significantly reducing the environmental footprint associated with traditional mining operations (Truby 2018). By utilising excess energy from renewable sources, these regions can both mitigate environmental harm and create new economic opportunities. Another example comes from New York City, where fintech-enabled energy trading platforms allow building owners to trade surplus renewable energy generated from rooftop solar panels. This decentralised energy market fosters collaboration between urban residents and reduces dependence on fossil fuels, enhancing the city's sustainability efforts (Tando et al. 2025). Looking forward, the convergence of sustainability and fintech could pave the way for new forms of economic resilience and environmental stewardship, especially in energy-rich regions. As emerging technologies like AI and quantum computing continue to develop, their integration into fintech could further amplify its benefits. AI is already making waves by improving efficiency, personalisation, and security in financial services. AI offers tailored products to customers based on their behaviour and preferences while also enhancing fraud detection systems. However, using AI has its challenges. Concerns about algorithmic bias and the risk of enabling socio-economic inequalities have been raised, making it clear that the ethical use of AI in fintech is not something to ignore (D'Acunto et al. 2019). In regional transportation planning, AI-driven fintech platforms are being employed to manage mobility-as-a-service (MaaS) models. For example, Helsinki's Whim application integrates payments for multiple transport modes, including public transit, bike-sharing, and ride-hailing services, streamlining urban mobility while reducing the need for private car ownership. This demonstrates how fintech can align with urban sustainability goals (Future Mobility Finland 2017).

Quantum computing, although it is still in its infancy, holds promise for transforming financial modelling and data analysis by enabling systems to process vast amounts of data at unprecedented speeds. Quantum computing could revolutionise areas such as risk management and portfolio optimisation, allowing fintech companies to make better informed decisions more quickly. However, it also raises security concerns, as quantum computers could potentially break existing encryption methods, necessitating the development of more advanced cryptographic techniques (Nielsen and Chuang 2010). In urban planning, quantum computing could enhance smart city simulations by analysing billions of variables, such as traffic flow, energy consumption, and emergency response scenarios, to design resilient cities optimised for sustainability and efficiency.

This chapter aims to explore the diverse potential of these technologies in promoting equitable and sustainable development within cities and regions. By examining the crucial role of fintech in bridging financial gaps and enhancing resource management, we can gain valuable insights into how these technologies can contribute to building more equitable and sustainable societies. Furthermore, the chapter situates fintech as a critical enabler of urban and regional transformation, offering practical solutions to pressing challenges such as resource efficiency, environmental degradation, and socio-economic inequality.

1.1 The Role of Fintech in Financial Inclusion

Access to financial services has traditionally been restricted by several barriers, including geographic isolation, inadequate infrastructure, and high transaction costs. These challenges are especially pronounced in rural and low-income areas. Operating in such regions is expensive, and the returns for banks are typically low, meaning large segments of the population remain financially excluded. Fintech has revolutionised this model by offering cost-effective solutions that do not require physical bank branches. Digital platforms provide a way for underserved populations to access services through mobile devices, which drastically lowers the cost of banking for both the institution and the consumer.

In many developing countries, mobile phones have become crucial for financial inclusion. For example, Kenya's M-Pesa allows users to deposit, withdraw, and transfer money using mobile phone networks (Suri and Jack 2016). The impact of mobile money systems like M-Pesa has been fundamental, especially for marginalised groups such as rural populations and women. These services have not only improved access to financial services but have also played a key role in reducing poverty by providing new ways to save, invest, and manage money (Demombynes and Thegeya 2012). Mobile money platforms allow users to perform transactions without the need for a bank account, reducing reliance on cash and increasing financial stability. Platforms such as Branch and Tala have taken this further by using alternative data, such as mobile phone usage and social media activity, to assess creditworthiness. This enables them to provide loans to individuals without formal credit histories, increasing access to financial services (Berg et al. 2020).

1.2 The Role of Digital Currencies in Financial Inclusion

Digital currencies and blockchain technologies are another frontier for financial inclusion. Blockchain, with its decentralised nature, enables secure, peer-to-peer transactions without the need for liaisons like banks or mobile operators. This is particularly useful in regions where banking infrastructure is underdeveloped or too expensive for most people. By allowing individuals to transact directly, blockchain reduces costs and increases the speed of financial transactions. This is especially beneficial for cross-border remittances, where fees and delays are often a significant burden. For migrant workers, digital currencies offer a cost-effective way to send money home, ensuring that more of their earnings reach their families (Foley et al. 2019).

Dubai's Blockchain Strategy exemplifies how blockchain can extend beyond financial transactions to address urban governance challenges. By digitising processes such as land registries and business licensing, Dubai has improved transparency and operational efficiency, offering a model for blockchain integration in municipal governance to achieve sustainability and accountability goals (Khan et al. 2022).

However, using digital currencies has many challenges. The decentralised nature of blockchain, while removing intermediaries, makes it difficult to regulate. This opens the door to potential misuse for illegal activities, such as money laundering or the financing of terrorism (Böhme et al. 2015). Furthermore, the technological complexity of managing digital wallets and understanding blockchain mechanics can act as barriers for those with limited financial and digital literacy. Moreover, in regions with limited internet availability or limited access to smartphones, the benefits of digital currencies may be out of reach, potentially widening the digital divide.

2 The Impact of AI on Fintech

AI is rapidly changing the fintech landscape, transforming how financial services are optimised and secured. At the heart of this transformation is AI's ability to process and analyse vast amounts of data in real time, allowing fintech platforms to offer customised financial products and services. AI is also streamlining processes, improving customer service, and providing more secure environments for financial transactions. This revolution in fintech is not just about technology; it is reshaping how individuals, especially those from underserved communities, interact with financial services. For instance, AI-powered robo-advisors have democratised wealth management. Traditionally, wealth management services were only available to high-net-worth individuals, but platforms like Betterment and Wealthfront are changing that by offering low-cost, automated financial advice. These platforms use AI algorithms to analyse an individual's financial data, assess their risk tolerance, and recommend personalised investment strategies. This means that people with limited financial literacy or smaller portfolios can access the kind of tailored financial advice previously reserved for the wealthy (D'Acunto et al. 2019). By providing this kind of personalisation, AI bridges the gap between underserved populations and traditional financial institutions, allowing more people to make informed decisions about savings, investments, and retirement planning.

Beyond wealth management, AI is also transforming lending and credit assessments. Traditional credit scoring systems rely on a narrow set of data, such as credit history and income levels, often leaving people without formal financial records unable to access credit. AI-driven platforms, however, are changing that. By analysing alternative data sources, such as mobile phone usage, transaction patterns, and social media activity, AI systems can offer a more comprehensive assessment of an individual's creditworthiness. This allows fintech companies to extend loans to individuals who may lack formal credit histories but demonstrate responsible financial behaviour in other ways. This expansion of credit access is particularly important for low-income populations and those in developing countries, where traditional banking services are often inaccessible (Berg et al. 2020).

Helsinki's Whim application demonstrates the potential of AI in regional transportation systems. By unifying payments for public transit, bike-sharing, and ride-hailing services, Whim streamlines urban mobility while reducing dependence on private vehicles, contributing to urban sustainability and efficient resource use (Kamargianni and Matyas 2017). This demonstrates the potential impact of AI on urban sustainability through the streamlining of financial services.

2.1 AI and Fraud Detection

Fraud detection is another area where AI is making significant contributions to fintech. Given the digital nature of fintech platforms, they are vulnerable to cyberattacks and fraudulent activity. AI helps mitigate these risks by using machine learning algorithms to detect anomalies in transaction data. These systems can flag suspicious activity in real time, preventing fraud before it escalates. By analysing the history of transactions, AI can detect patterns that indicate fraudulent behaviour, making it more effective than traditional fraud detection systems (Ngai et al. 2011).

AI can also enhance security through biometric authentication systems, such as facial recognition and voice authentication. These systems add an extra layer of protection for users, making it harder for unauthorised individuals to access accounts or conduct fraudulent transactions. In addition to fraud detection, AI-driven chatbots powered by natural language processing (NLP) are improving customer service. These AI tools handle routine inquiries and provide real-time support, ensuring that customer issues are resolved quickly and efficiently. As a result, fintech platforms can offer a more secure and user-friendly experience, which is crucial for maintaining customer trust in a digital financial environment.

2.2 AI's Role in Sustainable Financial Practices

AI can play an important role in promoting sustainability within fintech. One of the biggest areas where AI contributes to sustainability is optimising energy consumption in data centres and blockchain networks. Cryptocurrency mining and blockchain operations are notoriously energy-intensive, but AI-powered systems can predict usage patterns and adjust resource allocation to minimise energy waste. By monitoring and optimising these processes, AI helps reduce the environmental impact of these operations, making fintech platforms more sustainable, especially in regions with limited energy resources (OECD 2021).

AI is also driving the development of green financial products. Many fintech companies are now offering eco-friendly investment portfolios, carbon offset programmes, and other sustainable financial products. AI algorithms can analyse environmental, social, and governance (ESG) data to identify companies and investments that align with sustainability goals. By integrating these metrics into the investment process, AI empowers people and institutions to make more environmentally friendly financial decisions. This shift towards sustainability is part of a broader trend within fintech, which is to balance profitability with environmental responsibility. For example, some AI-driven platforms are helping investors build portfolios focused on companies with strong ESG performance. These tools allow investors to support businesses that prioritise sustainability, from reducing carbon footprints to practising responsible corporate governance. By automating the analysis of vast datasets related to environmental and social impacts, AI makes it easier for users to align their financial goals with their values.

Furthermore, AI is being leveraged to optimise energy usage within financial institutions themselves. Smart algorithms monitor energy consumption in fintech offices, data centres, and operations, identifying areas where energy can be saved. This not only reduces operational costs but also lessens the organisation's overall carbon footprint. AI systems that adjust lighting, heating, and cooling based on real-time usage patterns are just one example of how AI technology is driving energy efficiency within fintech operations.

3 Quantum Computing and Its Implications for Fintech

Quantum computing is set to revolutionise many industries, including fintech, by vastly improving how information is processed. Unlike classical computers that use binary bits (0s and 1s) to process data, quantum computers use quantum bits, or qubits, which can exist in multiple states simultaneously due to quantum superposition. This allows quantum computers to perform calculations at an exponentially faster rate than traditional systems (Nielsen and Chuang 2010). Financial markets generate massive amounts of data, and analysing this data in real time demands enormous computational resources. Quantum computing offers the capability to process such data efficiently, enabling faster and more accurate predictions of market trends, risks, and investment opportunities. This technological leap could give fintech companies a competitive edge in areas such as portfolio optimisation, trading, and algorithmic risk management (Orús et al. 2019).

One of the most critical applications of quantum computing in fintech is cryptography, which is crucial for ensuring the security of digital transactions and data. Traditional cryptographic methods, such as RSA encryption, rely on the difficulty of factoring large prime numbers. However, quantum computers could potentially crack these methods by solving complex problems that are beyond the reach of classical computers (Shor 1994). This poses a significant threat to current encryption standards, as quantum computers could render them obsolete.

Quantum cryptography provides a solution through Quantum Key Distribution (QKD), which relies on the principles of quantum mechanics to create encryption keys that are practically unbreakable. Unlike classical cryptography, QKD uses the inherent properties of quantum particles, ensuring that any attempt to intercept or tamper with the key would be immediately detectable, as observing quantum particles alters their state. By adopting quantum cryptography, fintech companies can safeguard sensitive financial data from future cyber threats posed by quantum computing advancements (Bennett and Brassard 2014).

3.1 Quantum Computing and Financial Modelling

Another promising application of quantum computing in fintech is its potential to revolutionise financial modelling and risk analysis. Financial institutions depend on complex models to predict market behaviour, assess risk, and optimise portfolios. These models often require solving complicated mathematical problems, such as simulating financial markets under various conditions or determining the likelihood of specific outcomes. Quantum computers have the ability to solve these problems much more efficiently than traditional computers, allowing fintech companies to create comprehensive financial models (Orús et al. 2019). For instance, quantum computers could optimise investment portfolios by simultaneously analysing multiple variables and identifying the best asset allocation strategies. Further, quantum computing could enhance risk management by allowing companies to simulate a broader range of market scenarios and stress-test financial systems under risky conditions. This ability to foresee potential vulnerabilities would help fintech companies develop more robust risk mitigation strategies, ultimately making financial markets more resilient to unpredictable events.

4 Challenges and Ethical Considerations in the Use of Advanced Technologies in Fintech

The integration of advanced technologies such as AI, quantum computing, and blockchain in fintech holds transformative potential. However, with such innovations come critical ethical and regulatory concerns. These emerging technologies promise to make financial systems more efficient and accessible, but they also risk exacerbating existing inequalities and introducing new ethical challenges. Balancing innovation with responsibility will require careful attention to issues such as bias, discrimination, data privacy, sustainability, and widening digital divide.

AI has rapidly become an integral part of fintech, particularly in automating processes like credit assessments, fraud detection, and personalised financial advice. However, the use of AI in credit assessment presents significant ethical challenges, particularly around bias and discrimination. AI systems are trained in historical data, which may carry biases that reflect existing socio-economic inequalities. If left unchecked, these biases can lead to discriminatory outcomes, disproportionately disadvantaging marginalised communities. For example, AI algorithms used in credit scoring often rely on a combination of financial and non-financial data, such as social media activity and online behaviour. If these models are not carefully designed, they can reinforce existing inequalities, particularly for people from low-income backgrounds who may lack a traditional credit history. Studies have shown that certain AI models used in lending decisions tend to discriminate against women or minorities, even when there is no overt intent to do so (Binns 2018). The problem lies in the training data; if the data is twisted, so too will the outcomes.

AI bias and discrimination have become more complex in a global context. AI models developed in one region may fail to account for the cultural and socio-economic differences of another. A model trained in the USA, for instance, may not perform equitably when applied to financial services in developing countries, where different financial behaviours, transaction histories, and economic realities exist. Addressing these biases will require diverse datasets and a commitment to fairness and inclusivity in algorithmic design. Moreover, transparency and accountability are crucial. Currently, many AI systems operate as "black boxes," where the decision-making process is opaque. This lack of transparency creates issues when customers or regulators try to challenge unfair lending or risk assessment decisions. Ethical AI in fintech demands more transparent and explainable algorithms that allow users to understand how decisions are made and offer recourse in cases of discrimination.

Quantum computing is poised to revolutionise fintech by radically improving data processing speeds, encryption, and financial modelling. While the benefits of quantum computing are undeniable, its adoption also presents ethical concerns, particularly around the digital divide. Quantum computing requires highly specialised resources, and initially, its benefits will likely be concentrated in large financial institutions that can afford the technology. Smaller fintech firms, especially in developing regions, may struggle to keep up, leading to a widening gap in technological and security competences (Orús et al. 2019).

One of the most profound impacts of quantum computing will be on encryption. Quantum computers have the potential to break current encryption standards, rendering traditional security measures outdated. This poses a major security threat, as sensitive financial data, personal information, and transactions could be compromised. Large institutions may be able to transition to quantum-proof encryption, but smaller institutions may not have the resources to do so, making them more vulnerable to cyberattacks. This technological gap could aggravate global inequalities as wealthier nations and large corporations gain access to quantum technologies while smaller businesses and developing countries are left behind struggling. To address these issues, policymakers will need to create strategies that promote equitable access to quantum technologies and ensure that the benefits of quantum computing are distributed across different sectors and regions. Moreover, ethical frameworks for the use of quantum computing in financial services must be developed to protect data privacy and hinder misuse.

Blockchain technology offers fintech decentralisation, independence, and a transparent way of conducting financial transactions. By removing the need for intermediaries, blockchain promises to make financial systems more efficient, secure, and accessible. However, it also raises ethical concerns, particularly around governance and regulation. One of the core benefits of blockchain is its decentralisation, but this also makes regulatory oversight challenging. Traditional financial systems are governed by a set of rules and institutions that ensure accountability, consumer protection, and market stability. In contrast, decentralised finance (DeFi) platforms operate without a central authority, which can make it difficult to enforce regulations. This lack of oversight poses risks for consumers, as fraudulent activities or market manipulation may go unchecked (Zohar 2015). Moreover, while blockchain increases financial access for many, its complexity and reliance on digital literacy can create barriers for others. Managing digital wallets, securing private keys, and understanding how blockchain transactions work require a level of technical knowledge that many individuals in developing regions may not have. This could further exacerbate the digital divide, as those with access to education and technology reap the benefits of blockchain while those who do not have this type of access are left behind.

The environmental impact of blockchain, particularly in cryptocurrency mining, is another significant ethical concern. The energy-intensive nature of proof-of-work consensus mechanisms, which underpin many cryptocurrencies like Bitcoin, has led to growing criticism over the sustainability of blockchain. Mining operations consume vast amounts of electricity, often sourced from fossil fuels, contributing to carbon emissions and environmental degradation (Mora et al. 2018). While there are efforts to develop more sustainable blockchain technologies, such as proof-of-stake systems, the ethical question remains: Can the financial benefits of blockchain justify its environmental cost?

4.1 The Sustainability of Advanced Technologies in Fintech

As fintech nowadays is adopting cutting-edge technologies such as AI, quantum computing, and blockchain, sustainability has become a growing concern. While these innovations offer immense potential for transforming financial systems, their environmental, social, and economic implications demand careful consideration. Addressing the sustainability of fintech is not just about mitigating its environmental impact but also about ensuring that technological progress fosters inclusivity and social responsibility.

Cryptocurrency mining, especially for Bitcoin, has become a focal point of environmental criticism. Mining, which relies on the proof-of-work (PoW) model, includes solving complex puzzles that require substantial computing power. This computational dependence translates into significant energy consumption. Bitcoin mining alone consumes more energy annually than entire countries like Switzerland (Mora et al. 2018). The high energy demand of Bitcoin mining is largely fuelled by non-renewable resources in many parts of the world, contributing to carbon emissions and environmental degradation.

At the heart of the controversy is the growing disparity between the benefits of blockchain and its environmental toll. Bitcoin mining, for instance, produces an enormous carbon footprint. Research shows that Bitcoin's energy consumption could push global warming beyond the 2 °C threshold set by the Paris Agreement if left unrestrained (Mora et al. 2018). With more people and companies investing in cryptocurrencies, the environmental impact is expected to rise unless more sustainable methods are adopted. There is, however, a movement towards some changes. For example, alternative consensus mechanisms like proof-of-stake (PoS) are gaining traction as more energy-efficient models. PoS eliminates the need for energy-hungry mining by allowing validators to confirm transactions based on the number of coins they hold, rather than their computational power. Ethereum, one of the leading blockchain platforms, is currently transitioning to PoS, which promises to reduce its energy consumption significantly (Beekhuizen 2021). Furthermore, beyond shifting to PoS, some regions with abundant renewable energy sources, such as Iceland and Norway, are exploring ways to power cryptocurrency mining sustainably. The Brooklyn Microgrid highlights how blockchain technology can facilitate peer-to-peer energy trading, empowering residents to buy, sell, and share renewable energy directly. This decentralised energy system fosters local resilience, reduces reliance on traditional power grids, and promotes the use of renewable energy sources (Mengelkamp et al. 2018). These countries leverage geothermal and hydroelectric energy to power mining operations, creating an opportunity for crypto to go green while still driving economic value (Truby 2018). These initiatives reflect a growing recognition within the crypto community that its environmental cost must be addressed head-on.

4.2 *AI and Energy Efficiency: A Double-Edged Sword*

AI is at the forefront of fintech innovation, driving efficiencies in everything from fraud detection to personalised financial products. However, the energy cost of AI has emerged as an overlooked sustainability issue. Training large AI models, especially deep learning algorithms, requires vast computational resources. These models rely on data centres that consume significant amounts of electricity. Indeed, research shows that training a single AI model can generate as much carbon dioxide as five gasoline-powered cars over their entire lifetimes (Strubell et al. 2020). While AI's energy demands are concerning, the technology itself holds promise for reducing the environmental footprint of fintech. AI-driven tools can optimise energy usage in data centres, helping to minimise waste and improve overall efficiency. By predicting peak demand periods, AI can dynamically adjust the allocation of computational resources, ensuring that energy is used as efficiently as possible.

AI can facilitate the creation of green financial products. Through the analysis of environmental, social, and governance (ESG) data, AI can help fintech companies guide their customers towards more sustainable investment options. Platforms powered by AI can offer personalised recommendations that align with consumers' environmental values, making sustainability a core part of financial decision-making (OECD 2021). Despite these positive applications, the sustainability of AI itself requires more attention. The rapid expansion of AI technologies across industries has prompted calls for more responsible energy management in data centres and for AI systems to be developed with sustainability in mind. Striking a balance between AI's energy demands and its potential to drive greener solutions will be critical as technology continues to evolve.

4.3 Quantum Computing and Sustainability

From a sustainability perspective, quantum computing offers some promising advantages. Quantum computers, when fully developed, could drastically reduce the energy required for complex financial computations. For instance, tasks such as optimising investment portfolios, which are currently energy-intensive, could be completed much more efficiently, reducing the overall energy consumption of fintech platforms. Moreover, quantum computing could help fintech companies enhance encryption without the massive energy costs associated with current data security systems (Orús et al. 2019). However, as with any new technology, the infrastructure required to support quantum computing presents its own set of sustainability challenges. Quantum computers need specialised environments, including extremely low temperatures, to function. Developing the quantum data centres necessary to house these machines could offset many of the sustainability benefits if they are not powered by renewable energy sources. Ensuring that quantum infrastructure is built with sustainability in mind will be crucial to minimising its environmental footprint.

4.4 Socio-economic Sustainability and Inclusivity

Beyond environmental concerns, fintech should address the broader social implications of these technologies. As advanced systems like AI, blockchain, and quantum computing become more integral to the financial industry, the risk of widening the digital divide grows. While large financial institutions and developed nations have the resources to invest in these cutting-edge technologies, smaller fintech firms and developing countries may struggle to keep pace. This could lead to an unequal distribution of technological benefits, where wealthier regions enjoy greater security, efficiency, and access, while others fall behind. For example, quantum computing promises to enhance encryption and financial modelling, but the costs associated with building and maintaining quantum infrastructure are likely to be prohibitive for smaller institutions. This creates the potential for a financial landscape where only the wealthiest firms have access to the most advanced technology, leaving smaller players vulnerable to cyber threats and unable to compete on equal terms (Orús et al. 2019).

Similarly, while AI and blockchain can expand access to financial services in underserved regions, they also require high levels of digital literacy and infrastructure. For these technologies to be truly sustainable, they must be deployed in a way that promotes inclusivity, ensuring that all communities, regardless of their technological capabilities, can benefit. Public–private partnerships that invest in digital infrastructure and digital literacy training in developing regions will be essential to achieving this goal.

5 Conclusion

In reflecting on the broader implications of fintech's role in shaping the future of cities and regions, its potential to foster more equitable and sustainable societies within urban and regional contexts must be emphasised. Each application of fintech—ranging from financial services to digital currencies, fraud detection, and energy optimisation—contributes uniquely to addressing urban challenges while aligning with the goals of sustainability and inclusivity. Fintech disrupts traditional financial models by using digital technologies to extend critical services to underserved populations. From mobile money platforms that empower rural communities to blockchain innovations that reduce remittance costs, these advancements mark a fundamental shift in how financial systems operate. In urban contexts, these innovations are particularly significant in reducing financial exclusion among low-income residents, improving access to essential services, and creating opportunities for local economic growth.

Sustainability is increasingly central to fintech's role in global development. The energy-intensive processes of cryptocurrency mining have drawn deserved scrutiny, with critics pointing out their significant contributions to carbon emissions. Therefore, some regions have begun to explore more responsible models by using surplus renewable energy for power mining operations. Urban ecosystems have embraced blockchain-driven energy trading platforms, such as the Brooklyn Microgrid, which exemplifies how fintech can integrate renewable energy into city-level resource management. By harnessing geothermal and hydroelectric energy to fuel blockchain activities, fintech creates a symbiotic relationship between financial innovation and environmental stewardship, supporting urban sustainability goals.

Furthermore, the integration of AI in fintech offers a dual promise. On the one hand, AI enhances the personalisation of financial services, enabling tailored solutions for users regardless of their socio-economic status. On the other hand, it can be a tool for environmental management, optimising energy use in data centres and reducing the carbon footprint of fintech operations. In smart cities, AI applications like Helsinki's Whim platform streamline urban mobility by unifying transportation payments and reducing reliance on private vehicles, fostering sustainable and inclusive transportation systems. Using AI technologies, however, has its ethical implications. Issues of algorithmic bias and transparency must be tackled head-on, ensuring that AI-driven fintech platforms serve all communities fairly and equitably. Urban planners must prioritise transparency and accountability in AI deployment to ensure that these technologies do not inadvertently reinforce inequalities within urban ecosystems.

Quantum computing has the potential to revolutionise fintech. Quantum computing performs complex calculations at speeds far beyond what classical computers can achieve, promising enhanced financial modelling and risk management. In the context of urban finance, quantum computing could support advanced simulations for resource allocation, risk assessment, and smart city planning, allowing municipalities to make data-driven decisions with greater precision.

However, this technology must be harnessed in ways that do not deepen the global digital divide. Smaller firms and developing regions may struggle to access these advancements, and without intentional strategies to democratise quantum technologies, fintech risks exacerbating global inequalities rather than alleviating them. Urban policymakers, therefore, must consider how quantum computing innovations can be integrated equitably, ensuring accessibility for both developed and developing regions.

The future of fintech is as much about responsibility as it is about innovation. Its role in urban finance and smart city ecosystems underscores the need to design systems that are not only efficient but also equitable and sustainable. From enabling decentralised energy trading to providing inclusive financial solutions and enhancing urban mobility, fintech's potential rests on its alignment with broader urban sustainability goals. To build a truly inclusive and sustainable future, industry must navigate a complex set of challenges, from ensuring equitable and inclusive access to managing the environmental costs of digital technologies. Financial institutions, policymakers, and technologists must work together to create an ecosystem where innovation thrives, but not at the expense of fairness or sustainability. Fintech has the capacity to foster economic resilience, but only if it does so with an eye towards both equity and sustainability. Ultimately, fintech has the capacity to transform urban and regional systems into more resilient, inclusive, and sustainable frameworks, paving the way for societies that balance technological advancement with equity and environmental stewardship.

Declaration of Competing Interests The authors have no conflict of interest.

References

Beekhuizen D (2021) Ethereum 2.0 and proof of stake: a new era for blockchain

Bennett CH, Brassard G (2014) Quantum cryptography: public key distribution and coin tossing. Theor Comput Sci 560:7–11

Berg T, Burg V, Gombović A, Puri M (2020) On the rise of fintechs: credit scoring using digital footprints. Rev Financ Stud 33(7):2845–2897

Binns R (2018) Fairness in machine learning: lessons from political philosophy. In: Conference on fairness, accountability and transparency

Böhme R, Christin N, Edelman B, Moore T (2015) Bitcoin: economics, technology, and governance. J Econ Perspect 29(2):213–238

D'Acunto F, Prabhala N, Rossi AG (2019) The promises and pitfalls of robo-advising. Rev Financ Stud 32(5):1983–2020

Demirguc-Kunt A, Klapper L, Singer D, Ansar S (2018) The Global Findex Database 2017: measuring financial inclusion and the fintech revolution. World Bank Publications

Demombynes G, Thegeya A (2012) Kenya's mobile revolution and the promise of mobile savings. World Bank policy research working paper (5988)

Digital Dubai (2020) Smart Dubai. Dubai Blockchain strategy. Retrieved 26 Nov 2024 from https://www.digitaldubai.ae/initiatives/blockchain

Foley S, Karlsen JR, Putniņš TJ (2019) Sex, drugs, and bitcoin: how much illegal activity is financed through cryptocurrencies? Rev Financ Stud 32(5):1798–1853

Future Mobility Finland (2017) Retrieved 26 Nov 2024 from https://futuremobilityfinland.fi/vision/
Kamargianni M, Matyas M (2017) The business ecosystem of mobility-as-a-service. Transportation Research Board
Khan S, Shael M, Majdalawieh M, Nizamuddin N, Nicho M (2022) Blockchain for governments: the case of the Dubai Government. Sustainability 14(11):6576
Mengelkamp E, Gärttner J, Rock K, Kessler S, Orsini L, Weinhardt C (2018) Designing microgrid energy markets: a case study: the Brooklyn microgrid. Appl Energy 210:870–880
Mora C, Rollins RL, Taladay K, Kantar MB, Chock MK, Shimada M, Franklin EC (2018) Bitcoin emissions alone could push global warming above 2 °C. Nat Clim Chang 8(11):931–933
Ngai EW, Hu Y, Wong YH, Chen Y, Sun X (2011) The application of data mining techniques in financial fraud detection: a classification framework and an academic review of literature. Decis Support Syst 50(3):559–569
Nielsen MA, Chuang IL (2010) Quantum computation and quantum information. Cambridge University Press
OECD (2021) AI and the environment. https://oecd.ai/en/footprint
Orús R, Mugel S, Lizaso E (2019) Quantum computing for finance: overview and prospects. Rev Phys 4:100028
Shor PW (1994) Algorithms for quantum computation: discrete logarithms and factoring. In: Proceedings 35th annual symposium on foundations of computer science
Strubell E, Ganesh A, McCallum A (2020) Energy and policy considerations for modern deep learning research. In: Proceedings of the AAAI conference on artificial intelligence
Suri T, Jack W (2016) The long-run poverty and gender impacts of mobile money. Science 354(6317):1288–1292
Tando MS, Gawusu S, Yakubu S, Kwabena OS, Donatus D, Abu SR (2025) Decentralized energy trading with blockchain technology. In: The intersection of blockchain and energy trading. Elsevier, pp 75–101
Truby J (2018) Decarbonizing bitcoin: law and policy choices for reducing the energy consumption of blockchain technologies and digital currencies. Energy Res Soc Sci 44:399–410
Zohar A (2015) Bitcoin: under the hood. Commun ACM 58(9):104–113

The Role of Insurance and Insurtech in Enhancing Natural Hazard Resilience

Paul Hudson

1 Introduction and Key Challenges Facing Natural Hazard Insurance

Climate change is one of the most pressing issues that society faces due to the multifaceted avenues through which the consequences of climate change are experienced. IPCC (2022) states that globally we will most likely see global warming of 1.5 °C in the period 2021–2040, compared to the current observed level of 1.1 °C over the last decade. The rapid warming of the globe will have significant knock-on effects on human wellbeing. There will be, among others, changes to agricultural productivity (Allanson et al. 2023), there will be negative impacts on human health (Watts et al. 2017), and there will be changes to the current observed patterns of natural hazards (Clarke et al. 2022). The change in global temperatures is now manifesting through changes in weather patterns and the increasing occurrence of extreme weather events, a statement that can now be made due to improvements in attribution science (Clarke et al. 2022). The increase in extreme weather events is particularly problematic due to the catastrophic impacts that such events can create. For example, Munich Re estimates that in 2022, USD 270 billion was lost due to natural hazard events, about 45% of which was not insured (Munich Re 2023). This has grown from USD 162 billion in 2013 of which only 26% was insured (Munich Re 2023). These are truly significant impacts that as a society we must all work towards managing and limiting. This is particularly required because it is often recognised that in society it is those who are most vulnerable that bear the most disproportionate impacts of climate change and natural hazard events (Parsons et al. 2024) and that overall, the threat posed by natural hazards will only worsen (IPCC 2022).

Therefore, there has been a greater focus on integrated disaster risk management or resilience-building efforts and policy in work that looks at social vulnerability,

P. Hudson (✉)
Department of Environment and Geography, University of York, York, UK
e-mail: paul.hudson@york.ac.uk

M. Okitasari et al. (eds.), *Equitable and Sustainable Cities*, Advances in 21st Century Human Settlements, https://doi.org/10.1007/978-981-92-2793-8_10

adaptation, and resilience research (Kuhlicke et al. 2023, 2020). There are many competing perceptions of resilience and how it should be defined. However, for disaster risk reduction, the concept is defined in the Sendai Framework for Disaster Risk Reduction 2015–2030 (UNISDR 2015) as "*The ability of a system, community or society exposed to hazards to resist, absorb, accommodate to and recover from the effects of a hazard in a timely and efficient manner, including through the preservation and restoration of its essential basic structures and functions*" a viable definition to use. While resilience is a complex topic with multiple levels of action, facilitators and enablers, a key aspect of building resilience is financial inclusion and capacity (Mhlanga 2022). Financial inclusion is key because a significant aspect to resilience is the ability to finance activities that allow people impacted by natural hazards to resist impacts or recover from impacts in an affordable manner (Valencia et al. 2024). Therefore, people will need access to suitable financial mechanisms so that these costs can either be spread over time [e.g. through loans such as in Hudson (2020)] or over many geographically diversified people to share the burden [e.g. through insurance such as in Hudson et al. (2019)]. Insurance as a financial mechanism is a prominent element of policy in this area because high levels of insurance coverage, in theory, mean that when a disaster event occurs there can be a rapid influx of planned finances allowing those impacted to get back on their feet quickly and limit indirect and long-term consequences, while also providing potential incentives for additional risk management and risk reduction activities by the policyholder either due to insurance-based incentives or as conditions for coverage (Hudson et al. 2016). The ex-ante nature of insurance reduces the pressure of ex-post actions and financing which by its nature may be of a more ad-hoc nature (Unterberger et al. 2019).

However, while insurance has a great number of positive attributes for disaster and natural hazard resilience in the face of climate change, there are significant challenges (Charpentier 2008). One is that there is an underinsurance problem across the globe (Holzheu and Turner 2018). Swiss Re (2023) estimates that, for example, the global crop protection gap at USD 113 billion, while about 75% of global natural catastrophe risk was unprotected in 2022, with protection gaps largest in emerging markets. Underinsurance occurs either because people are not buying insurance or when they are not buying sufficient insurance (e.g. their property has a reconstruction cost of GBP 100,000 but they have only purchased GBP 50,000 of coverage). While there can be rational reasons for underinsurance to a degree (e.g. a deductible to lower the premium and limit moral hazard), it can be considered as a type of financial exclusion. Financial exclusion or inclusion has been defined in several different ways by different authors. However, the World Bank (2022) defines financial inclusion as "*the process where individuals and businesses have access to useful and affordable financial products and services that meet their needs like transactions, payments, savings, credit and insurance.*" Therefore, limited access to affordable products would be considered a financial exclusion.

The ability of the insurance industry to provide affordable products is determined by the conditions of insurability. The conditions of insurability can be defined as following Charpentier (2008):

1. There is a sufficient degree of occurrence randomness rather than certainty.
2. The possible maximum loss should be within the insurer's solvency capacity.
3. The impact should be identifiable and quantifiable.
4. Risks should be pooled and, mostly, be independent in their claims for portfolio diversification and the annual expected loss can be calculated.
5. There should be non-systematic or limited moral hazard (i.e. no negative behavioural changes once insured), and adverse selection (i.e. no hidden or misleading information on the risk level)
6. There must exist sufficient overlap between the price that insurers are willing-to-accept and buyers willing-to-pay for a given insurance product for a market to exist.

These conditions are at times difficult to achieve for natural hazard insurance. For example, adverse selection is a potential issue [see, e.g. Hudson et al. (2017)] as we see that because of a tendency to underperceive the threat posed by low-probability events that only those most threatened want to buy insurance, which in turn limits condition 4 holding because of the geographical clustering of insurance demand. Moreover, conditions 2 and 6 can also be difficult to meet because of informational barriers (e.g. no existing flood hazard maps meaning that bespoke hydrological models must be developed) or because of transaction costs (e.g. that new claims infrastructure and processes must be created in order to provide the product), which can increase the price that insurers need to sell at in order for it to be a commercially viable product, thereby reducing the overlap with what customers are willing to pay. Therefore, where the conditions of insurability are difficult to achieve, we may see limitations on where coverage is provided [e.g. see the withdrawal of private sector hurricane insurance in Florida (Kunreuther and Lyster 2016)] or that products are unaffordable because of the growing risk level, costs of provision, and increased liquid capital reserves required to remain solvent. In this vein, climate change will lead to an increased frequency and intensity of climate and weather-based disasters, which will lead to a greater degree of loss exposure and uncertainty for insurers who will charge higher premiums, lessen the coverage they are willing to provide, or withdraw entirely (Sheehan et al. 2023). This has been seen with (forest) fire insurance in California as well as windstorm insurance in Florida, which led to the creation of Citizens Property Insurance Corporation to act as an insurer of last resort.

Therefore, the problem of underinsurance and the pressures of growing insurability can be seen to be a problem of growing financial exclusion. In certain jurisdictions, this has been addressed with new insurance structures, for example, the introduction of Flood Re in the UK to create an insurance pool for the highest-risk properties to prevent the removal of the universal private sector flood insurance coverage (Surminski 2018). In other areas, where insurance coverage is seen to be technically difficult to provide, there is a push to provide parametric or index insurance products because of their less complex nature, which lowers the transaction costs of entering these markets at an affordable price (Collier et al. 2009).

These are pressures that must be overcome for insurance to remain a suitable avenue for building social resilience to natural hazards, rather than a tool that generates and/or exacerbates social inequalities (Lucas and Booth 2020). One supply-side approach to trying to maintain insurance capacity is through the use, and leveraging of novel financial technology (Fintech) and its subset of insurance technology (Insurtech) to make the most of novel and emerging technologies to mitigate or attenuate the increasing pressures, we see being placed on disaster insurance. For example, see ongoing discussions surrounding the ability of blockchain to support the development of smart contracts in parametric insurance products so that they more rapidly distribute resources in an open and transparent manner (Sandland et al. 2019), or how artificial intelligence (AI) or machine learning (ML) when combined with big data can be used to improve risk modelling and identification in areas of data scarcity or high complexity (Huntingford et al. 2019), or how generative AI could be used to process contracts or process claims more effectively and efficiently (EIOPA 2021), or the potential of the internet of things (IoT) to, within reason, keep track of the property-level precautionary activities of the policyholder or to act as a single/incentive or behavioural change (Spender et al. 2019) or improve insurance processes (Liu et al. 2022).

The ability of novel technologies to create new business models for affordable and accessible insurance can be demonstrated via the growth in parametric insurance. The rationale behind parametric insurance is that we often see marginalised communities disproportionately impacted by climate change due to limited resources for adaptation and recovery (IPCC 2022). To address the need to facilitate their recovery process, parametric insurance has emerged as a low-cost solution, offering small-scale risk coverage in exchange for small-scale premiums (Kousky et al. 2021; Mose 2022). These policies, with strict maximum payout limits, ensure immediate access to capital in the event of a natural hazard event, and these insurance policies can be offered as parametric or index insurance (Kousky et al. 2021; Sheehan et al. 2023). These parametric or index-based policies provide immediate pre-agreed upon payouts upon verification of adverse climate events, following a pre-agreed threshold or trigger being reached, using spatial and temporal monitoring tools like local meteorological, hydrological, and climatological resources, drones, and satellite imaging (Lin and Kwon 2020). Parametric insurance offers significant benefits due to its flexibility and simplicity in product design and risk accommodation. Unlike indemnity insurance, parametric insurance allows the insured to use the claim payment for any recovery purposes, offering additional freedom. This flexibility reduces disputes over covered risks and enables the insurance to adapt to various coverage periods aligned with specific risks.

The ability of Insurtech to operate, support, and change multiple aspects of the insurance industry across both traditional indemnity and parametric insurance means that it has a great potential to disrupt the insurance industry (Allanson et al. 2023) and may have a substantial impact on the ability of insurance to build climate and disaster resilience if we consider how these various and rapidly innovating examples of Insurtech fit within the social context in which they interact with natural hazard insurance. The remainder of this chapter is focussed upon providing a perspective on how Insurtech can be leveraged to build resilience.

2 Potential Impact of Insurtech on Natural Hazard Resilience

2.1 Defining Natural Hazard Resilience

Natural hazard resilience is a complex topic that is difficult to define within a single conceptual framework, as can be seen by the large number of such frameworks developed (Béné and Doyen 2018; IPCC 2022). However, one such framework is that resilience can be divided into three pillars, which consist of: resistance, recovery, and adaptive capacity (Hudson et al. 2020; Thieken et al. 2014). While other frameworks [e.g. that of Béné and Doyen (2018) or de Graaf-van Dinther and Ovink (2021)] can produce additional pillars of resilience, it can be argued that this is often done by breaking down one of the three pillars into what could be considered its subcomponents. For example, adaptive capacity could be broken down into components relating to potential learning within current system vs. the ability to transform the system. Therefore, Béné and Doyen (2018) rightly consider conceptualisations of resilience as being along a spectrum of considerations that choose highly specific aspects of resilience to be prioritised.

In the proposed conceptualisation of resilience, the three pillars can be defined as follows. Resistance is the ability of the individual or system to withstand and limit the impacts of the natural hazard directly or indirectly. Recovery is the ability of the system to return to the pre-hazard state and overcome the burdens placed on it by the natural hazard event. Adaptive capacity is the ability to learn, improve, and transform such that we can progressively move forwards proactively to lessen the impacts of the next natural hazard event.

Insurtech can interact with these different pillars of resilience through different entry points that even provide new opportunities to the insurance industry and potential policyholders or allow them to become more efficient at their current processes.

2.2 The Resistance Pillar of Natural Hazard Resilience

One of the most common arguments for how insurance coverage develops resilience's resistance pillar is by the implementation of additional risk reduction by the policyholder (Kuhlicke et al. 2020; Kunreuther and Lyster 2016). This is because while insurance itself doesn't lower the impacts of a disaster, insurance coverage only redistributes the costs of disasters, it can provide rewards for the policyholder to implement adaptive measures such as wet or dry flood-proofing (Hudson et al. 2016). This is because with a lower level of risk, the premium should fall, incentivising the policyholder to reduce the threat faced where it is cost-beneficial to do so (Hudson et al. 2016). However, despite the widespread call for such activities to be proactively rewarded by insurers, there is little systematic evidence of their presence outside of deductibles.[1] One argument put forward for this is the presence of transaction costs in monitoring the behaviour of the policyholder to make sure that the risk is reduced (Hudson et al. 2016). The monitoring potential of intermittent drone imagery may be one way of lowering these transaction costs. A second possible avenue for reducing the transaction costs that inhibit insurance-based incentives for risk reduction is the Internet of Things (IoT). IoT refers to a network of interconnected physical devices embedded with sensors and software, enabling them to collect and exchange data over the Internet. The applicability of IoT will differ depending on the measure but potentially offers a way to monitor the success and implementation of property-level risk reduction measures. Kopetz and Steiner (2022) note that the true value of the IoT is not in the actual connectivity of the devices themselves but in the fact that IoT-enabled or potential devices and monitors have become very widespread and ubiquitous. The IoT could generate a wealth of data that helps offset the information asymmetries between the insurer and the insured that are the root cause of the absence of mass incentives for policyholder-level risk reduction.

2.3 *The Recovery Pillar of Natural Hazard Resilience*

Maximising the recovery potential of policyholders can be argued to be the central objective of insurance coverage for resilience building. The potential for rapid, predictable, and reliable compensation payments is how insurance builds resilience. However, despite this being the core mechanism and business of insurance, there are still limitations for resilience via insurance. For instance, in the case of a large event rapid insurance pay-outs may not be possible due to the assessment and claims process. For instance, this was a problematic experience after the Christchurch earthquake in 2011 and Hurricane Katrina in 2005 where payments were staggered over many years (Corey and Deitch 2011; Green et al. 2007; Poontirakul et al. 2017; Potter et al. 2015). Poontirakul et al. (2017) note that adequate and prompt payments promote recovery, while slow payment processes may be no better than receiving no compensation payments at all as was found in the case of those impacted by the Christchurch earthquake in 2011.

Therefore, the first way that novel technologies can aid in insurance boosting resilience is via improvements in claims management, customer interaction, and company processes. For example, parametric insurance is often argued to speed up and simplify the process of making, processing, and paying out claims, rendering

[1] A deductible is a condition of an insurance product where the policyholder must cover a pre-set amount of any claim before the insurer will payout the remainder of the claim.

insurance more accessible to those requiring coverage due to the simplified processes and clear rules and conditions (Kousky et al. 2021; Lin and Kwon 2020; Sandland et al. 2019). The processes behind insurance contracts could be further supported with novel technologies. For instance, smart contracts on blockchain platforms can automate payouts when specific conditions are met, such as a weather event of a sufficient magnitude or length, following clearly defined rules. Smart contracts on a blockchain can automate claims payments once predefined criteria are met, based on real-time data inputs from catastrophe monitoring technologies. This automation significantly reduces administrative burdens, speeds up capital deployment, and mitigates human errors (Braun et al. 2021; Shetty et al. 2022).

Supporting traditional insurance policies in such a manner can reduce costs through increased automation. Moreover, these novel technologies can enable accurate recording, reporting, and measuring in a publicly accessible and tamper-proof manner. Transparent documentation of experiences, claims, and agreements also aids in generating larger datasets for big data analysis that can provide new insights into changing patterns of risk. However, such advantages can be limited in the case of parametric insurance because of the limited way in which strategic actions by the policyholder can influence insurance payouts or premiums. Though traditional indemnity insurance may see such benefits manifest if multiple sources of information are linked into the blockchain record (e.g. IoT records of the implementation and status of property-level flood resilience measures).

Additionally, the ability of insurance to promote faster post-event recovery is based on the wide accessibility of insurance coverage. Improving insurance accessibility across various markets, particularly in the context of disaster-related risks, presents numerous challenges. For example, in low-income markets, insurance coverage can be low because a suitable financial payment infrastructure is not present (Sandland et al. 2019). The lack of an accessible financial infrastructure can render insurance non-accessible because the initial investment required will render the product non-profitable. Therefore, we have seen that improvements in technology can "leapfrog" traditional approaches to providing insurance, thereby creating new business models that can sidestep the current barriers to insurance provision. For example, the development of a robust mobile payment infrastructure can significantly lower transaction costs, making various insurance types more accessible because the insurance policy can fit into the technology and processes that the customer is familiar with. Moreover, policies profitable for middle-class consumers are often unscalable for "base of the pyramid" customers due to high transaction costs associated with customer acquisition, policy issuance, record maintenance, and claim settlement. In these markets, digital and mobile communication are transformative, dramatically reducing the required investment costs to work with this segment of the market, thereby facilitating insurance provision and accessibility.

2.4 The Adaptive Capacity Pillar of Natural Hazard Resilience

Adaptive capacity is a broad conceptual component of resilience, one element of which can be considered as how widely accessible natural hazard insurance coverage is to those at risk. Widespread insurance coverage can bring a range of benefits as not only can individuals recover faster if businesses are also insured, this can lessen indirect economic impacts as well (Koks and Thissen 2016). Therefore, the ability of novel Insurtech to transform the way insurance is provided or operates is a key part of social adaptive capacity and building resilience by closing the insurance protection gap (The Geneva Association 2016).

Within this vein, the continued or starting provision of disaster insurance is a key priority for policymakers (Hudson et al. 2020) as the aforementioned conditions for insurability must be recognised so that the industry can be future proofed as a viable component of disaster resilience (Sheehan et al. 2023). Insurtech can potentially operate in this space to assist in overcoming the threat posed by underinsurance because of the potential that new technology can change both the demand and supply aspects of natural hazard insurance. The fundamental source of underinsurance is the perceived affordability of insurance products and the willingness of the insurer to provide coverage, which are underpinned by technical issues such as risk estimation, costs of capital provision, etc. which can all be impacted by novel technologies.

Advances in Insurtech could significantly impact the affordability of insurance that is economically advantageous. This is because novel aspects of Insurtech can address some of the features that lead to higher premiums. For example, a common source of higher premiums is the transaction costs of observing, monitoring, and providing a natural hazard insurance product and managing the claims process after a hazard event. The integration of big data and AI by insurance companies into their customer and claims management process can increase the efficiency through which customer accounts are managed. For example, consider the use of chatbots and their potential interaction with large language models (LLM) to operate within the narrow bounds of an insurance transaction and claim processes[2] without direct human involvement in most cases. Moreover, dependent on the scale of the insurance company the integration of big data and AI, across a range of datasets, into the actuarial process may increase the accuracy at which products can be priced either because uncertainty in the natural hazard process is reduced or behavioural traits such as adverse selection or moral hazard can be more readily identified. However, unlike the reduction in transaction costs the influence of improved risk assessment and profiling is heterogeneous. This is because the insurer can more reliably segregate policyholders meaning that some will receive lower premiums, while others get higher premiums.

A further way in which large data and AI can help process the information the insurance industry needs is via improvements in satellite and drone imagery. These

---

[2] Though this may be easier with parametric insurance vs. traditional indemnity insurance.

novel technologies represent alternative data streams providing real-time views on risk exposure and impacts, which further necessitates the application of advanced machine vision techniques for analysis so that the quantity of data can be managed successfully, creating a required cycle of investments in data science and engineering infrastructure within an insurance company.

The successful integration of these novel technologies into the actuarial process can offer, in principle, multiple ways of lowering insurance premiums. One source of transaction costs within the natural hazard insurance sector is the development of the natural hazard risk profile. This is expensive because in many locations across the world, historical data doesn't have a long enough time series to provide a reflection of the threat, doesn't have a suitable level of accuracy or quality, or due to climate change, is no longer reflective of the underlying natural processes. Therefore, there has been a rapid growth in catastrophe modelling combined with climate projections to model potential impacts. These models are often regionally, nationally, or globally focussed due to the data they require. While they can be based on land-use data and globally available meteorological data for weather-based hazards, this does introduce an element of measurement error into the risk estimates for a given property holder. Therefore, drone technology and improvements in imaging equipment that can be placed on a drone can allow for autonomously inspecting potentially insured properties. Drones can create accurate 3D models of properties, analysed through machine vision processing algorithms, and automatically detect natural hazard risks a property may face, such as fire risks due to nearby fuel loads, by analysing present materials and their properties. This capability would enable insurance companies to better estimate actual risks, reducing the cost of expert inspections and enhancing the affordability of insurance products. This will be especially valuable with growth in open science and open access models and data that further reduce the upfront investment costs required for risk assessment and evaluation.

3 Critical Perspective on the Role of Novel Technologies for Future Natural Hazard Insurance

While novel insurance technologies discussed above offer great potential for increasing the affordability and accessibility of insurance, in turn allowing for a higher level of resilience, they also present several challenges and risks that must be addressed to ensure the continued provision of natural hazard insurance in the face of climate change and current institutional arrangements. This is because the adoption of technologies such as blockchain introduces concerns about performance scalability, energy consumption, data privacy, and integration with legacy systems.

One relates to the transparency and immutability of blockchain, which can conflict with data protection laws like GDPR, particularly regarding the right to be forgotten. Similarly, systems based on integrating the IoT with insurance policies and potential risk reduction incentives will generate a large amount of information and potential cyber-attacks further raise large concerns with data protection and security, especially if data is used to profile individuals into different risk or behaviour segments to base insurance business decisions upon. Regulatory and legal uncertainties compound these issues (Jia 2023). The novel technologies that can underpin such revised business models and activities can operate in a legal grey area, with varying regulatory frameworks across jurisdictions. This lack of uniformity can impede the widespread adoption and reliability of Insurtech solutions for natural hazard insurance. This could be particularly problematic for reinsurers who act as a global aggregator for natural hazard risk. Additionally, EIOPA (2021) emphasises the need to implement robust data governance processes when using datasets generated by novel technologies to protect public privacy and foster trust in insurance.

Additionally, the concern around security is particularly concerning in regions most vulnerable to natural hazards, where technological infrastructure may be lacking. Moreover, novel applications of technologies such as the IoT can require secure and consistent connectivity across and between objects as well as the insurer receiving this information must, in turn, have the capacity to understand and act upon this information in a socially useful way. The ability of the insurer to act upon this information is also potentially counterproductive from a social point of view. For example, the use of AI to identify patterns in large datasets of policyholders can lead to increasing risk segmentation as new, unintended, ways of identifying different risk levels are seen. While increasing risk segmentation would be in favour of the general insurability conditions, it implies a reduction in risk pooling.

The model of insurance developed by Rothschild and Stiglitz (1978) implies that in the presence of information asymmetries on risk levels, we can reach a market equilibrium where insurance coverage is offered jointly to high and low-risk actors if we are not able to sufficiently distinguish between risk levels but the willingness-to-pay of high and low-risk actors sufficiently overlap. The benefit of this is that an implicit cross-subsidy is generated which maintains the affordability of natural hazard and climate risk insurance for those most at risk. However, if we can segment higher and lower-risk individuals more precisely, we will see premiums diverge between the two groups. Those at higher risk will see their premiums increase, while those at lower risk will see their premiums reduced. Hudson et al. (2019) find that insurance systems will lower degrees of risk pooling and see a reduction of insurance affordability and penetration rates in the areas at the highest risk. This can be socially counterproductive because the individuals forgoing insurance are possibly the ones who require insurance the most. The implications of this will differ across countries with systems that embrace solidarity as the leading factor, such as France or Spain, or countries where risk differentiation is more important such as Germany. The negative consequences might be more clearly felt in cases like Germany where private insurers seek to differentiate between risk levels to increase their profitability and solvency, while increasing risk segmentation might be advantageous in the case of France because capital reserves can be more accurately planned for as well as guiding national investments in risk reduction infrastructure and property-level incentives and communication strategies.

An additional potential concern is that the novel technologies introduced to reduce the costs of providing insurance could rather than promoting access could see insurance become a more rigid mechanism. For example, smart contracts are an ideal match with parametric insurance due to the ability of parametric insurance to be expressed as a series of "if/or" statements that can be verified with either publicly available data or data collected via the IoT. We may see increasing issues with managing contingencies and consumer complaints, which are critical in the dynamic environment of natural hazard impacts. Moreover, the complexity of these technologies may widen the gap between consumer understanding and product features, potentially leading to misinterpretations and inadequate consumer protection. This could further reduce the level of trust that consumers have in the insurance industry due to increasing perceptions of "inhumanity" of the process, which could worsen the demand problems related to basis risk.

While smart contracts may benefit from blockchain's transparency, this might not be true of other uses of novel technologies such as AI that can be used to manage processes or underwrite risks. The inherent complexity of novel technologies can lead to a lack of transparency for consumers because the insurance industry increasingly works with "black boxes." Understanding the workings of blockchain-based insurance or the algorithms behind AI-driven risk assessments requires a high level of technical knowledge, which most consumers lack. This complexity can create misunderstandings about insurance products and undermine consumer trust. For the insurance industry, consumer trust and acceptance are critical to maintain a sufficient consumer basis if purchase decisions are not legally compelled.

Additionally, Dominguez Anguiano and Parte (2024) note that key insurance stakeholders are still sceptical about how and when blockchain or AI might bring value to their businesses outside of a high-level perspective. Similarly, Jia (2023) notes that the value of blockchain has yet to be empirically demonstrated by insurance which limits the incentive for private profit-driven insurers to embrace unproved and potentially unprofitable technologies.

Environmental and energy impacts of these technologies cannot be ignored. Technologies like blockchain and AI are known for their high energy consumption, which can be counterproductive in the context of climate change. For example, if the energy consumed is generated by fossil fuels, then this worsens the potential degree of climate change experienced along with a worsening disaster risk profile. This counterproductive relationship might be mitigated if such technologies were powered with renewable energy sources for example. However, this does not diminish the significant development and operating costs that these technologies generate. While these costs may fall in the future as the technologies develop and refine, their use in the short term may come at the expense of a concentrated insurance market as only certain insurance companies can bear the costs in return for a sustainable competitive advantage. A reduction in market competition may also be counterproductive for insurance systems that rely on the private market because without competitive pressures insurance companies will not have as strong an incentive to pass on the benefits of novel technologies to consumers.

4 Conclusion

Disaster insurance is a key element of global resilience to disasters, as not all disasters can be prevented nor all damage successfully mitigated. Insurance and insurance-like mechanisms have an important role to play in making sure those impacted by a disaster event have an influx of resources. However, insurance systems themselves are complex processes integrating several information chains that, in turn, sit at the end of a long chain of decisions that generate and manage climate risk. Therefore, there are multiple opportunities for how novel technologies can be deployed in the insurance sector at different points along these two processes that can be leveraged directly or indirectly to produce a higher level of climate resilience.

The use of novel technologies in the insurance sector can help build resilience because they can help to reduce the "gap" between the insurance company and its customers allowing for a more productive dialogue, while also allowing the insurance company access to a wider range of "big data" sources that can be used to optimise the insurance companies' business and underwriting processes into a more competitive position. Moreover, as there has been a very rapid growth, unequally across the globe, in smart devices that can both provide and receive data and notifications, we may see opportunities to use the IoT to overcome, or reduce, the problems surrounding transaction actions and information asymmetries that currently inhibit the mass implementation of insurance-based incentives for additional risk reduction activities by the policyholders. The technologies that enable these movements, therefore, provide opportunities to enhance the ability of insurance to act across three pillars of resilience: We can see an improvement in the ability to resist climate risk through more individualised risk reduction incentives; we can see an increased ability of insurance to aid in recovery because technical innovations could reduce the costs of providing insurance and therefore more generous or affordable insurance in competitive markets, as well as speeding up the claims process so that recovery aid is more timely; we can also see improvements in adaptive capacity because novel technologies allow us to make greater sense of the wealth of data that insurers generate both in terms of understanding risk and the responses of their policyholders and as such new opportunities for building climate risk resilience.

However, despite the potential advantages provided by novel insurance technologies, there are several caveats and potential limitations that must also be kept in mind. The first is that there needs to be a robust regulatory framework that establishes suitable oversight and standards on the use and generation of data to maintain suitable security and protection. While establishing this framework will be troublesome given the rate at which novel technologies have developed over the last few years, while it is possible that advancements in this area may plateau, regulation will be unlikely to suitably track developments given the much slower pace of regulation, legislation, and political processes. The second caveat is that the development of these novel technologies can have unintended consequences, such as a concentration of insurers if novel technologies generate a sustainable competitive advantage given the development costs or a reduction in insurance coverage because consumers can be more easily risk-segmented, which can render insurance unaffordable to those most at risk and as such lower coverage where it is most needed. The final caveat to be made is that the call for the development and implementation of novel technologies as the solution to climate and disaster resilience could evolve into a rhetorical point like the call for better data and better use of data. This can be seen as a way of delaying action that can be done today because we are waiting for "the solution" to manifest through a rather passive process of generating and using improved flows of information and knowledge. While there is an element of truth to these pervasive calls, we must also note as Mami Mizutori does in her introduction to the mid-term report on the Sendai Framework: "*However, at the midpoint of the implementation of the 2015 agreements, progress has stalled and, in some cases, reversed. This has resulted not only from the impact of the Covid-19 pandemic, but also from short- versus long-termism, weakened multilateralism, disconnects between the real and the financial economies, rising inequality, and barriers between risk science, perception and risk-informed decision-making.*" Therefore, we should remember that while the use and generation of novel technologies can aid in improving resilience, we must avoid this becoming a call akin to creating a future parallel reality that is better rather than undertaking and challenging the socio-political drivers of vulnerability and fragility and the range of actions that can be undertaken today.

Declaration of Competing Interests There are no competing interests to report.

References

Allanson E, Combina A, Richelli P, Zaharieva L (2023) The United Kingdom: the nexus of Insurtech

Béné C, Doyen L (2018) From resistance to transformation: a generic metric of resilience through viability. Earth's Fut 6(7):979–996. https://doi.org/10.1002/2017EF000660

Braun A, Haeusle N, Karpischek S (2021) Decentralized autonomous risk transfer on the blockchain. In: The routledge handbook of FinTech. Routledge, pp 69–78

Charpentier A (2008) Insurability of climate risks. The Geneva papers on risk and insurance—issues and practice 33(1):91–109. https://doi.org/10.1057/palgrave.gpp.2510155

Clarke B, Otto F, Stuart-Smith R, Harrington L (2022) Extreme weather impacts of climate change: an attribution perspective. Environ Res Clim 1(1):012001

Collier B, Skees J, Barnett B (2009) Weather index insurance and climate change: opportunities and challenges in lower income countries. Geneva Pap Risk Insur Issues Pract 34:401–424

Corey CM, Deitch EA (2011) Factors affecting business recovery immediately after Hurricane Katrina. J Conting Crisis Manage 19(3):169–181. https://doi.org/10.1111/j.1468-5973.2011.00642.x

de Graaf-van Dinther R, Ovink H (2021) The five pillars of climate resilience. In: de Graaf-van Dinther R (ed) Climate Resilient Urban Areas: governance, design and development in coastal delta cities. Springer International Publishing, pp 1–19. https://doi.org/10.1007/978-3-030-57537-3_1

Dominguez Anguiano T, Parte L (2024) The state of art, opportunities and challenges of blockchain in the insurance industry: a systematic literature review. Manage Rev Q 74(2):1097–1118. https://doi.org/10.1007/s11301-023-00328-6

EIOPA (2021) Artificial intelligence governance principles: towards ethical and trustworthy artificial intelligence in the European insurance sector.

Green R, Bates LK, Smyth A (2007) Impediments to recovery in New Orleans' upper and lower ninth ward: one year after hurricane Katrina. Disasters 31(4):311–335
Holzheu T, Turner G (2018) The natural catastrophe protection gap: measurement, root causes and ways of addressing underinsurance for extreme events†. Geneva Pap Risk Insur Issues Pract 43(1):37–71. https://doi.org/10.1057/s41288-017-0075-y
Hudson P (2020) The affordability of flood risk property-level adaptation measures. Risk Anal 40(6):1151–1167
Hudson P, Botzen WJW, Aerts JCJH (2019) Flood insurance arrangements in the European Union for future flood risk under climate and socioeconomic change. Glob Environ Change 58:101966. https://doi.org/10.1016/j.gloenvcha.2019.101966
Hudson P, Botzen WJW, Feyen L, Aerts JCJH (2016) Incentivising flood risk adaptation through risk based insurance premiums: trade-offs between affordability and risk reduction. Ecol Econ 125:1–13. https://doi.org/10.1016/j.ecolecon.2016.01.015
Hudson P, Botzen WW, Czajkowski J, Kreibich H (2017) Moral hazard in natural disaster insurance markets: empirical evidence from Germany and the United States. Land Econ 93(2):179–208
Hudson P, De Ruig LT, de Ruiter MC, Kuik OJ, Botzen WJW, Le Den X, Persson M, Benoist A, Nielsen CN (2020) An assessment of best practices of extreme weather insurance and directions for a more resilient society. Environ Hazards 19(3):301–321. https://doi.org/10.1080/17477891.2019.1608148
Huntingford C, Jeffers ES, Bonsall MB, Christensen HM, Lees T, Yang H (2019) Machine learning and artificial intelligence to aid climate change research and preparedness. Environ Res Lett 14(12):124007
IPCC (2022) Climate change 2022: impacts, adaptation, and vulnerability. In: Pörtner HO, Roberts DC, Tignor M, Poloczanska ES, Mintenbeck K, Alegría A, Craig M, Langsdorf S, Löschke S, Möller V, Okem A, Rama B (eds) Contribution of working group II to the sixth assessment report of the intergovernmental panel on climate change. Cambridge University Press
Jia R (2023) Assessing the potential of decentralised finance and blockchain technology in insurance
Koks E, Thissen M (2016) A multiregional impact assessment model for disaster analysis. Econ Syst Res 28(4):429–449
Kopetz H, Steiner W (2022) Internet of things. In: Kopetz H, Steiner W (eds) Real-time systems: design principles for distributed embedded applications. Springer, pp 325–341. https://doi.org/10.1007/978-3-031-11992-7_13
Kousky C, Wiley H, Shabman L (2021) Can parametric microinsurance improve the financial resilience of low-income households in the United States? Econ Disast Clim Change 5(3):301–327. https://doi.org/10.1007/s41885-021-00088-1
Kuhlicke C, Madruga de Brito M, Bartkowski B, Botzen W, Doğulu C, Han S, Hudson P, Nuray Karanci A, Klassert CJ, Otto D, Scolobig A, Moreno Soares T, Rufat S (2023) Spinning in circles? A systematic review on the role of theory in social vulnerability, resilience and adaptation research. Glob Environ Change 80:102672. https://doi.org/10.1016/j.gloenvcha.2023.102672
Kuhlicke C, Seebauer S, Hudson P, Begg C, Bubeck P, Dittmer C, Grothmann T, Heidenreich A, Kreibich H, Lorenz DF (2020) The behavioral turn in flood risk management, its assumptions and potential implications. Wiley Interdiscip Rev Water 7(3):e1418
Kunreuther H, Lyster R (2016) The role of public and private insurance in reducing losses from extreme weather events and disasters. Asia Pac J Environ Law 19(1):29–54
Lin X, Kwon WJ (2020) Application of parametric insurance in principle-compliant and innovative ways. Risk Manage Insur Rev 23(2):121–150. https://doi.org/10.1111/rmir.12146
Liu L, Li W, He W, Zhang JZ (2022) Improve enterprise knowledge management with internet of things: a case study from auto insurance industry. Knowl Manag Res Pract 20(1):58–72. https://doi.org/10.1080/14778238.2021.1970490
Lucas CH, Booth KI (2020) Privatizing climate adaptation: How insurance weakens solidaristic and collective disaster recovery. Wiley Interdiscip Rev Clim Change 11(6):e676
Mhlanga D (2022) The role of financial inclusion and FinTech in addressing climate-related challenges in the industry 4.0: lessons for sustainable development goals. Front Clim 4:949178

Mose S (2022) Microinsurance as a strategy in enhancing insurance penetration in Kenya University of Nairobi

Munich Re (2023) Natural disaster risks: rising trend in losses. https://www.munichre.com/en/risks/natural-disasters.html

Parsons ES, Jowell A, Veidis E, Barry M, Israni ST (2024) Climate change and inequality. Pediatr Res. https://doi.org/10.1038/s41390-024-03153-z

Poontirakul P, Brown C, Seville E, Vargo J, Noy I (2017) Insurance as a double-edged sword: quantitative evidence from the 2011 Christchurch earthquake. Geneva Pap Risk Insur Issues Pract 42:609–632

Potter SH, Becker JS, Johnston DM, Rossiter KP (2015) An overview of the impacts of the 2010–2011 Canterbury earthquakes. Int J Disaster Risk Reduc 14:6–14

Rothschild M, Stiglitz J (1978) Equilibrium in competitive insurance markets: an essay on the economics of imperfect information. In: Uncertainty in economics. Elsevier, pp 257–280

Sandland C, Schilling D, Marke A (2019) A critical analysis on the capacity of blockchain-based parametric insurance in tackling the financial impact of climatic disasters. In: Strengthening disaster resilience in small states: commonwealth perspectives, p 87

Sheehan B, Mullins M, Shannon D, McCullagh O (2023) On the benefits of insurance and disaster risk management integration for improved climate-related natural catastrophe resilience. Environ Syst Decis 43(4):639–648. https://doi.org/10.1007/s10669-023-09929-8

Shetty A, Shetty AD, Pai RY, Rao RR, Bhandary R, Shetty J, Nayak S, Keerthi Dinesh T, Dsouza KJ (2022) Block chain application in insurance services: a systematic review of the evidence. SAGE Open 12(1):21582440221079876. https://doi.org/10.1177/21582440221079877

Spender A, Bullen C, Altmann-Richer L, Cripps J, Duffy R, Falkous C, Farrell M, Horn T, Wigzell J, Yeap W (2019) Wearables and the internet of things: considerations for the life and health insurance industry. Br Actuar J 24:e22. https://doi.org/10.1017/S1357321719000072

Surminski S (2018) Fit for purpose and fit for the future? An evaluation of the UK's new flood reinsurance pool. Risk Manage Insur Rev 21(1):33–72

Swiss Re (2023) Restoring resilience

The Geneva Association (2016) Harnessing technology to narrow the insurance protection gap

Thieken A, Mariani S, Longfield S, Vanneuville W (2014) Preface: flood resilient communities–managing the consequences of flooding. Nat Hazards Earth Syst Sci 14(1):33–39

UNISDR (2015) Sendai framework for disaster risk reduction 2015–2030

Unterberger C, Hudson P, Botzen WJW, Schroeer K, Steininger KW (2019) Future public sector flood risk and risk sharing arrangements: an assessment for Austria. Ecol Econ 156:153–163. https://doi.org/10.1016/j.ecolecon.2018.09.019

Valencia DC, Calabuig C, Salazar JFL, Morales ME (2024) Digital financial inclusion as a pillar of disaster resilience. In Perez-Uribe R, Ocampo-Guzman D, Salcedo-Perez C, Carvajal-Contreras A (eds) Organizational management sustainability in VUCA contexts. IGI Global, pp 241–259. https://doi.org/10.4018/979-8-3693-0720-5.ch013

Watts N, Adger WN, Ayeb-Karlsson S, Bai Y, Byass P, Campbell-Lendrum D, Colbourn T, Cox P, Davies M, Depledge M (2017) The Lancet Countdown: tracking progress on health and climate change. Lancet 389(10074):1151–1164

World Bank (2022) Financial Inclusion. https://www.worldbank.org/en/topic/financialinclusion/overview#1

Investing in Co-Benefits: A Financial-Economic and Political Explanation

Jurgen van der Heijden, Shanita Rambharos, and Hade Dorst

1 Introduction

The concept of benefit-sharing has gained significant traction in recent years, particularly in the context of spatial development for sustainable cities and policy implementation. This phenomenon involves the co-financing of initiatives by parties who derive indirect benefits from the implementation of these initiatives. While traditional examples, such as road managers contributing to dike construction, are well-established, more complex scenarios have emerged, challenging conventional funding models. For instance, the potential role of health insurance companies in co-financing urban green spaces, which may lead to improved public health outcomes, exemplifies the intricacies of emerging benefit-sharing arrangements.

The integration of positive externalities, or co-benefits, into project planning and policy formulation represents somewhat of a paradigm shift in governance approaches. However, financial constraints often impede the realisation of integrated projects. This study aims to uncover the primary obstacle: the deeply entrenched vertical administrative culture governing expenditure (Andrews 2022; Mazzucato 2019; Scott and Gong 2021). Furthermore, it seeks to identify potential solutions, currently experimented within practice, to overcome this barrier. Our research synthesises several empirical studies within Dutch local governments and literature on financing integration, focussing on the governance of integrated projects and the emergence of novel financing strategies.

J. van der Heijden (✉) · S. Rambharos · H. Dorst
TNO, Vector, The Hague, The Netherlands
e-mail: jurgen.vanderheijden@tno.nl

S. Rambharos
e-mail: shanita.rambharos@tno.nl

H. Dorst
e-mail: hade.dorst@tno.nl

M. Okitasari et al. (eds.), *Equitable and Sustainable Cities*, Advances in 21st Century Human Settlements, https://doi.org/10.1007/978-981-92-2793-8_11

The term "beneficiary" has only recently been introduced to distinguish between parties that benefit from but do not contribute to a project, free riders, and those that can be engaged as potential co-financiers (Stipo and DML 2022). This distinction reflects a growing trend towards recognising and capitalising on co-benefits, moving away from a narrow focus on primary outcomes (Cohen-Shacham et al. 2016). The shift towards integration has prompted local governments to adopt new forms of cooperation, both within governmental organisations and between public and private sectors. This collaborative approach extends to the development of innovative financing strategies for integrated projects (Cumiskey et al. 2019; Duivenvoorden et al. 2021). However, this emerging horizontal mode of governance often conflicts with traditional, vertically oriented funding mechanisms characterised by predetermined and siloed budgets and rigid control mechanisms (Mazzucato 2019; Stipo and DML 2022).

Vertical governance typically refers to hierarchical structures where decision-making authority is centralised, often leading to siloed operations within governmental organisations. This model is characterised by rigid control mechanisms and predetermined budgets, which can hinder flexibility and responsiveness to emerging needs and collaborative opportunities (Scott 2020). In contrast, horizontal governance emphasises collaboration and partnership across various sectors, including public, private, and civil society actors. This model promotes integration and shared decision-making, facilitating the pooling of resources and knowledge to achieve common goals (Boström and Karlsson 2013).

The tension between vertical and horizontal governance models is particularly evident in the financing of integrated projects. This dynamic can be contextualised within the broader shift from government to governance (Stout and Love 2018). While integration represents the latest development in horizontal governance, it is clear that vertical governance structures persist, potentially leading to conflicts in project implementation and financing. Despite the growing recognition of co-benefits and the potential for smart project linkages (Floater et al. 2016), there remains a significant gap in the substantiation and financial underpinning of integrative strategies. Many local governments are already experimenting with integrated finance and co-financing projects (Stipo and DML 2022), but these efforts face numerous obstacles (Droste et al. 2017; Toxopeus and Polzin 2021).

To explore and analyse the current status of implementing integrative strategies in urban development projects, this study addresses the question: *How do public and private parties invest in integrated spatial development for sustainable cities, and what barriers must they overcome in this process?*

To answer the first part of this question, we conduct an analysis based on case study research supported by literature that examines new financial instruments and methods to support integration, which are currently utilised in the Dutch urban development context. To answer the second part of the question, we propose and test a theory of "trumping," wherein vertically grown governance culture and integrated solutions alternately prevail. We explore how financial, economic, and political arguments interact and potentially supersede one another in the context of integrated project financing for sustainable cities.

This chapter is structured as follows. First, we provide a theoretical framework that synthesises systems theory and action theory perspectives on integrated governance, culminating in the "theory of trumping." Next, the methodology section details the research design, including an overview of the empirical work and literature that was used for our analysis. Furthermore, we depict the framework used to identify emerging financial strategies and barriers. In the Results section, we present our analysis of emerging financial strategies for financing integrated spatial solutions. Using a case study analysis and the theory of trumping, we then present and reflect on possible opportunities and barriers to the implementation of novel financial strategies. Finally, we reflect on our findings and explore new research opportunities in our discussion in the future research section.

2 Theoretical Framework

2.1 Systems Theory Approach to Governance

Vertical coordination, characterised by hierarchical structures, is often deemed valuable due to its capacity for rapid decision-making and implementation. This form of coordination stands in contrast to horizontal coordination, which emphasises collaboration across different sectors. Despite the potential advantages of horizontal coordination, hierarchy often prevails as the default approach in governance, sometimes at the expense of more suitable alternatives (Scott 2020).

The tendency of governments to favour vertical coordination can be elucidated through the concept of "socio-technical regimes." These regimes consist of established norms and rules—institutional structures, such as political, legal, and financial constructs—that stabilise socio-technical systems (Fuenfschilling and Truffer 2014; Geels 2011). A socio-technical system, such as the energy system, represents a coherent network of actors and infrastructures providing specific societal functions. The stability of these regimes ensures the efficient functioning of the system but also constrains innovation and alternative organisational methods (Dorst et al. 2021).

The hierarchical and departmentalised nature of government organisations exemplifies regime norms, where vertical coordination is maximised to achieve efficient outcomes within silos (Scott 2020). These regimes can be so deeply entrenched that they become implicit and almost invisible, posing significant challenges to alternative working methods. Government departments tend to operate in isolation, limiting their interactions with other siloed departments, private organisations, and civil society organisations. Fragmentation within governance manifests through inadequate horizontal coordination, poor communication, and disputes over funding and responsibilities during the devolution of power to local authorities (Scott 2020).

While the development of market and network-based coordination mechanisms can mitigate the failures of silos, the ideal of vertical coordination often renders these solutions elusive (Scott 2020; Stipo and DML 2022). Silos persist due to their historical and cultural legacy as traditional, rational, bureaucratic forms of government. They enable politicians and officials to centralise political control, make rapid decisions, and implement effective economic strategies (Scott 2020).

The belief in the efficacy of vertical coordination assumes that hierarchical values can be realised in practice (Scott 2020). However, structural problems frequently obstruct this ideal. Financial or legal departments may veto execution, delaying decision-making processes. Information silos can lead to decisions based on incomplete data, and organisational units may become constrained by their own proliferating regulations intended to ensure employee compliance (Scott 2020).

Furthermore, the silo mentality, a significant cause of organisational failure, persists even in entities striving for collaborative work (Scott 2020). Interdepartmental competition for funding and bureaucratic control over policy processes further exacerbate this issue. Departments must secure funding for their activities, and success often hinges on political priorities and bureaucratic lobbying (Scott 2020).

Despite its frequent failures, vertical coordination remains a prevalent choice in governance, particularly within the dominant Dutch decision-making regime. Attempts at horizontal coordination repeatedly encounter resistance from this entrenched regime. When vertical coordination is chosen inappropriately, it forms a barrier to more efficient horizontal mechanisms, such as integration. This theorem, derived from Scott's work, underscores the need to explore solutions through action theory.

2.2 Action Theory Approach

In contrast to the systemic view, action theory focuses on the practical implementation of integration. Action theory is established on the conceptualisation of purposive behaviour through the lens of means-ends relationships. Spanning multiple disciplines, including economics, law, public administration, and business management, it posits a key starting point: Action is inherently goal-oriented, involving the strategic deployment of resources to achieve desired outcomes with optimal efficacy and efficiency.

However, an alternative perspective is emerging: the utilisation of singular means to accomplish multiple ends simultaneously. This approach challenges the traditional notion of efficiency, which typically emphasises maximising output while minimising input for a single objective. Puts and Van der Heijden (2017) and Revi et al. (2014) emphasise the importance of integrating multiple purposes into single actions, particularly in urban planning and environmental management. This multipurpose approach aligns with the concept of ecosystem services, where natural elements like trees can simultaneously provide multiple benefits such as pollution removal, cooling, and CO^2 storage (RLI 2018).

Building upon this alternative perspective, Van der Heijden and Vedder (2024) introduce the concept of "new efficiency," which refers to the capacity of a single action or intervention to achieve multiple objectives simultaneously. This notion expands the traditional understanding of efficiency by incorporating the potential for synergistic effects and multifaceted value creation. Consider residential housing as an example: While its primary function is to provide optimal living conditions, it can simultaneously serve additional purposes, such as healthcare provision, energy production, and climate adaptation. These secondary functions, or positive externalities, exemplify the potential for new efficiency. The synergistic interaction between these various functions can lead to a compounding of benefits, resulting in outcomes that exceed the sum of individual effects because of their integration.

2.3 A Theory of Trumping

In the context of integrated solutions, we propose a "theory of trumping" as a framework for understanding the dynamics between vertical and horizontal coordination in governance structures. This theoretical construct builds upon Scott's (2020) systems theory approach to governance and incorporates action theory to elucidate the interplay between maximisation and optimisation in decision-making processes.

Scott's systems theory approach delineates the gap between vertical and horizontal governance structures, providing a foundational concept of maximisation as a clear action goal. Maximisation aligns with vertical coordination, while optimisation aligns with horizontal coordination (Termeer and Metze 2019). This theoretical framework can thus complement Scott's systems theory by offering a practical application. In this context, optimisation is a state where no participant can improve their outcome without diminishing another's. This stands in contrast to maximisation, which seeks to achieve the highest possible outcome for a single party, often at the expense of others. A key tenet of the theory is that optimisation, associated with horizontal coordination, can trump maximisation, which is linked to vertical coordination. This occurs when parties recognise the long-term sustainability and mutual benefits of optimisation over the short-term gains of maximisation.

However, the theory also acknowledges the inherent gravitational pull of vertical coordination, especially within governmental structures where it is deeply ingrained. We propose that the appeal of vertical coordination (maximisation) and horizontal coordination (optimisation) coexist within governmental decision-making processes. While integrated, horizontal solutions may offer compelling arguments, the gravitational force of vertical solutions can often override these considerations. This phenomenon is attributed to the entrenched nature of hierarchical structures and the socio-technical regimes that stabilise existing systems (Fuenfschilling and Truffer 2014; Geels 2011).

To counter the dominance of vertical coordination, the theory suggests that new efficiencies or multiple value creation can provide significant resolving power. These factors can potentially convince stakeholders to opt for horizontal coordination even when vertical coordination would otherwise prevail. This aspect of the theory opens avenues for innovative financial thinking and the development of new instruments to support integrated solutions.

In conclusion, the theory of trumping offers an understanding of the tensions between vertical and horizontal coordination in governance. It synthesises insights from systems theory and action theory to explain how and why certain coordination mechanisms prevail over others; it helps explain why, despite the rational benefits of integrated approaches, vertical coordination often prevails in government decision-making. It also suggests that by emphasising new efficiencies and multiple value creation, proponents of integration can overcome the gravitational pull of traditional vertical structures.

3 Method: Synthesis of Practice Research and Analytical Focus

A series of practice-oriented (empiric) research projects on the development of new financial instruments and methods for the realisation of integrated projects have led us to reflections on why such integrations remain a challenge. This chapter employs a qualitative, multi-method approach to synthesise insights from various practice-oriented studies on integrated urban planning and financing. We combine case study analysis, document review, and semi-structured interviews to examine the challenges of implementing integrated projects and new financial instruments.

We introduce the "failed" case of the Cleantech region investment platform as an illustration of how the attractions and rational arguments of horizontal coordination and integrated investment still seem not compelling enough; it is trumped by other—and to the researchers still murky—arguments. As such, this case raises important questions regarding how decision-making around integrated planning is approached, and why seemingly rational arguments are surpassed or ignored. We further reflect on the theory of trumping, and to what extent and in what ways this takes place in practice, based on comparable instances of financing integration in practical research. Part of this research has been commissioned by Dutch governments in recent years as part of the City Deal Public Space (City Deal Openbare Ruimte or CDOR), and the other part is a series of interviews with representatives of seven Dutch cities.

3.1 Analytical Framework

The importance of integrating multiple objectives into single actions, particularly in urban planning and environmental management, was briefly discussed in the action theory approach section. In this chapter, we will further analyse the rationales and business cases for integrated solutions and joint investment strategies through the following aspects:

- Financing methods—Analysing practical approaches to joint investment in integrated interventions and preconditions for financing integration

- Governance and decision-making—Comparing vertical and horizontal governance structures to identify barriers to integration and potential solutions, and the arguments prevailing in the decision-making process

This framework guided the synthesis of findings across multiple data sources to develop an understanding of the challenges and opportunities in financing integrated spatial developments. The combination of case study, document analysis, and interview data allows for triangulation of findings and provides an empirical basis for theoretical reflection on the concept of "trumping" in decision-making processes around integrated planning initiatives.

3.2 Data Collection Through Practice Research

Two commissioned reports from the City Deal Public Space (*City Deal Openbare Ruimte*) were analysed:

- Stipo en Dark Matter Laboratories (DML) (2022), *Financiering van integratie - nieuwe ontwerpmethodiek en financieel instrumentarium voor de stedelijke transitie in de openbare ruimte*, Londen/Amsterdam 2022. www.platformisor.nl/artikelen/financiering-van-integratie
- De Natuurverdubbelaars (2024) *Investeren in integraliteit, Integrale fondsvorming voor meervoudige waardecreatie in de openbare ruimte.* City Deal Openbare Ruimte, www.platformisor.nl/artikelen/investeren-in-integraliteit

These reports were systematically reviewed to extract key themes and findings related to integrated financing methods and urban transitions.

In addition, semi-structured interviews were conducted with representatives from seven Dutch cities (Amsterdam, Eindhoven, Groningen, The Hague, Helmond, Rotterdam, and Utrecht) in spring 2023. These cities are participants in a two-year pilot project on neighbourhood-oriented investment platforms as part of the EU Cities Mission for climate-neutral cities (NetZeroCities 2024). The interview data was compiled into a report (not yet publicly available); these will be referred to as *Interviews NL cities*, which serve as a key data source for this study.

3.3 Illustration: Case Study Cleantech Region

A case study of the Cleantech region investment platform is presented as an illustrative example of the barriers to horizontal coordination and integrated investment approaches. The Cleantech region is a regional partnership of eight municipalities and two provinces in the east of the Netherlands, together representing 440,000 inhabitants and around 40,000 businesses. These governments are partnering with the relevant Dutch national ministries to enable societal and sustainable innovation focussing on the energy transition, circular economy, and human capital. This case—their initiative to establish an investment platform—was selected purposively to highlight how rational arguments for integration can be superseded by other factors in decision-making processes. Data for this case study was derived from three primary documents:

1. The Plan of Approach (PoA)
2. The Investment Manifesto 2.1
3. Memorandum on Building Blocks

These documents, while not publicly available, provide rich contextual information on the development and challenges of the investment platform initiative.

4 Results

4.1 Financing Methods

Traditional financial instruments are typically structured to support a single beneficiary's project, leveraging the capabilities of that beneficiary—a loan, for instance. Integration requires forms of financing in which other benefit holders also play a role, such as collaborative borrowing. To facilitate and formalise such arrangements, investment platforms and funds specifically designed for integration are emerging.

An investment platform is not a financial instrument like a loan; rather, it is a mechanism for organising financing. Similarly, a fund is not an instrument itself but can utilise instruments like loans to allocate resources. Both investment platforms and funds optimise financial flows and enable new channels of funding, fostering the cooperation and comprehensive investment necessary for integration.

Investment Platforms

Investment platforms function as brokers or matchmakers, organising project financing by facilitating interactions between financiers and projects. Making integrated investments is challenging without a platform that performs preliminary work, such as identifying synergies and trade-offs between societal values and seeking an optimal balance for both financiers and society. Knowledge sharing and building within the platform can further enhance this optimum. By sharing knowledge and costs, the involved parties reduce transaction costs and risks. It is also crucial to assess the risk of a party withdrawing by exploring how cooperation benefits them compared to acting independently. The smaller the difference, the higher the risk, and understanding the costs to bridge the gap is essential—the higher these costs, the greater the risk (source: Interviews NL cities).

An investment platform can be organised in various ways, such as a civil servant convening colleagues for a funding issue or a government-organised task force bringing actors together. Often, an investment platform develops business cases for projects, ensuring that ownership remains within a geographical area. The platform accelerates the creation of investable and financeable cases, meaning that investing and financing parties are convinced of the project's worth, due diligence has been conducted, and the expected and desired returns and impacts are sufficiently demonstrable (De Natuurverdubbelaars 2024).

Funds

Members of an investment platform collaborating on an integrated project contribute to a joint budget. Coordination is essential to fully benefit from cost reductions and enhanced quality achievable through collaboration. Yet there is a distinction between directing all funding flows separately to the integrated project and doing so in a structured manner through a collective fund. The latter approach, with a structured budget for the short and long term and various conditions, is more efficient. A fund can offer a solution for better and more responsible spending on an integrated project.

An investment platform can focus on project financing without forming a fund but can also serve as the preliminary phase of fund formation. Unlike a platform, a fund is a formally registered entity that is professionally managed. A fund can consolidate different financial flows from public and private parties, which can be jointly utilised to realise an integrated project. Practical examples of funds for integration are still scarce. Existing funds often have broad spending objectives but lack significant integration in the projects they finance.

It is crucial for participants to clearly understand a fund's purpose and the types of projects it supports. A holistic approach can dilute the fund's focus by attempting to achieve too many different values, obscuring its purpose and making it unclear to outsiders what the fund stands for. This lack of clarity can deter investors from participating in the fund. However, excessive focus conflicts with the desire for integration.

Pre-phase project funding is critical as integration can be incorporated into the plan during this phase. This phase often requires more time and money than single-goal projects. Parties must agree on the interpretation of projects and cost-sharing. The preliminary phase allows for the development of smart, integrated plans that pay off during the realisation and management phases. This phase can be organised and funded by an investment platform separately or become part of fund financing (De Natuurverdubbelaars 2024).

4.2 Governance Process and Decision-Making

Cleantech Region—A Case Illustration

In 2022, the Cleantech region proposed to establish a regional investment platform to facilitate integrated project development. This platform was intended primarily as a mechanism to address the limitations faced by individual organisations in the region, mobilising funding for projects that exceeded the risk tolerance of individual municipalities. As such, this would enable the realisation of large-scale initiatives such as heat grids, neighbourhood development, housing, and agricultural transition.

The investment platform would serve to bring together residents, governments, businesses, knowledge, and financial institutions to enhance project investment readiness. The platform was presented as a solution that could ensure that projects have an integrated effect, securing the connection of different policy themes. Additionally, the platform aimed to strengthen inter-stakeholder cooperation and facilitate the development of flexible business cases focussed on optimising broad prosperity, risk, and return, rather than solely maximising economic profit.

A key feature of the proposed platform was its ability to consider non-financial costs and long-term consequences of investment decisions. This approach allowed for the incorporation of avoided future societal costs into present-day investment decisions, even when these costs were not directly borne by the investing organisation. Moreover, the platform was positioned as a potential solution to address the diminishing resources available to decentralised authorities from the national government, which were increasingly insufficient for current tasks and left no margin for futureproofing.

The investment platform was envisioned to enable decentralised authorities to generate their own cash flow, retaining it in public hands for reinvestment in public initiatives. This model aimed to retain social value within the region and maintain regional control over critical infrastructures such as heat networks. It was argued that this approach would prevent the outflow of profits to private parties and external shareholders, thereby morally justifying the realisation of profits by public entities for social reinvestment.

Despite a generally positive reception, the executive committee of the Cleantech region did not approve the realisation of the investment platform. This decision may be attributed to several factors, including the novelty of integrated approaches for financiers and risk managers. Additionally, legal considerations regarding the permissibility of government profit-making likely influenced the decision.

The case highlights a preference for vertical governance and coordination approaches over horizontal integration, despite the apparent advantages of the latter. However, based on the available sources, there are too few insights into the potential arguments trumping those in favour of an integrated investment platform. Thereby, this case raises the question as to why, in a case where the option for horizontal integration seemed favourable, a vertical governance and coordination approach still came out on top. It underscores the need for further research into the decision-making processes of regional authorities, particularly in balancing innovative approaches with traditional governance structures and legal constraints.

Arguments and Trump Cards

The investment platform and the fund offer opportunities for financing integrated projects. Findings from the Cleantech region investment platform case show that integration is initially considered but ultimately not chosen, even if it appears to be the more advantageous option.

This case study mirrors the tension between vertical and horizontal coordination mechanisms that we describe in our theoretical framework. The findings demonstrate how economic and financial arguments—or trump cards—are leveraged to persuade politically accountable administrators to adopt an integrated solution. However, these arguments are ultimately superseded by political considerations that exercise a more substantial influence in this particular context. The case study illuminates the interplay of financial, economic, and political factors in decision-making processes related to integrated project financing; we recognise economic, financial, and political trump cards.

The Financial Trump Card

The investment platform in the Cleantech case study demonstrates significant financial advantages, primarily due to the pooling of resources which not only amplifies the budget but also creates sufficient trust to attract additional financial contributions, such as bank loans. Parties providing loan capital are encouraged to join the investment platform, increasing the total available funds, thereby ensuring the financial viability to implement large-scale projects. The inclusion of governmental organisations within the platform is particularly advantageous: The creditworthiness of the government is high, and it can take more risk than other parties; this can win over lenders to participate.

Commitment from parties to invest collectively is crucial. Binding contracts, as opposed to less formal covenants, solidify the relationships among co-investors, enabling long-term capital commitments. This contractual security can also attract risk-averse entities, such as debt providers, who may then be willing to offer loans at reduced interest rates. The investment platform fosters trust among potential co-investors, thereby facilitating these financial commitments (source: Interviews NL cities).

Similar to the investment platform in the Cleantech case, a dedicated fund could have contributed to increasing financial muscle and confidence. An additional advantage of a fund is its ability to finance a portfolio of projects, thereby distributing the risk of individual project failures across multiple investments. This risk diversification typically results in lower interest rates for the fund's financial contributions.

For instance, the initial investment required for a heat network is substantial. This scenario is often illustrated by the metaphor of an empty bathtub, representing the debt incurred for the network's construction. Over time, revenue gradually fills the bathtub, covering the incurred debts. However, the prolonged presence of debt accrues interest, necessitating strategies to expedite revenue generation.

One effective strategy to accelerate revenue generation is to leverage the increase in property values resulting from the heat network's installation. Beneficiaries of this increased value could waive their gains, providing collateral for the developer of the heat grid. This collateral allows the developer to secure relatively low-cost loans, thereby filling the bathtub more quickly (source: Interviews NL cities).

In sum, the financial trump card underscores the financial benefits of integration, facilitating the innovative use of both the investment platform and a dedicated fund.

The Economic Trump Card

The economic advantages of integration can be broken down into four principal arguments, which are illustrated with the examples of green roofs and solar panels. These are typical examples of applications benefitting sustainable cities.

Reduced Costs

One significant impact of green roofs is their ability to retain water, thereby allowing for disconnection from the sewer system. When a sufficient number of roofs are disconnected, it obviates the need for an expanded sewer system, resulting in substantial cost savings. Consequently, water authorities provide subsidies to property owners to incentivise the installation of green roofs, effectively sharing the construction costs. Although the initial investment in green roofs is higher compared to traditional black roofs, the long-term benefits include lower maintenance costs and extended roof lifespan, making them economically advantageous. Additionally, green roofs can serve as sites for solar panel installation, thereby saving scarce urban spaces and reducing associated costs at alternative locations (STOWA 2018).

Increased Yields

The presence of greenery on roofs contributes to neighbourhood cooling and enhances the living environment. Green roofs that retain rainwater offer even better cooling effects. Green roofs can also be combined with solar panels, cooling them, thereby ensuring more effective operation and increasing their energy output. However, the cooling effect on the neighbourhood may be diminished when solar panels are installed on green roofs. Nonetheless, green roofs exemplify multifunctional solutions that deliver additional quality benefits. For complex societal challenges such as climate adaptation, green roofs offer multiple advantages: They reduce energy consumption, mitigate heat stress, improve air quality, enhance biodiversity (especially with extensive green roofs), contribute to a quieter neighbourhood, and facilitate rainwater harvesting (STOWA 2018).

The additional quality of solar panels installed on green roofs translates into financial gains through electricity generation. For benefits like biodiversity and noise reduction, government entities can adopt a similar approach to water authorities when co-investing in green roofs. Monetising these benefits can justify public investment, potentially extending to values such as social cohesion, welfare, safety, and overall living environment quality. For instance, a water square can significantly enhance these benefits, providing a rationale for municipal co-financing from social budgets (source: Interviews NL cities).

Reduced Risk

Solar panels constructed over agricultural crops can protect them from extreme weather conditions such as heat, rain, and hail, thereby potentially increasing yields. Healthy soil and crops, in turn, help cool the panels, enhancing their electricity production. This reciprocal protection reduces risks for both farmers and power companies. Farmers benefit from reduced crop damage, while power companies mitigate the risk of panel overheating. Consequently, the cost of risk hedging, including borrowing and insurance, may decrease. However, it is also possible that lenders and insurers perceive integration as an increased risk, potentially raising their prices.

Transaction Costs and Real Options

Collaboration between farmers and energy companies incurs transaction costs, particularly during the construction phase. Continued coordination over time, which would not occur otherwise, implies additional transaction costs. Nevertheless, this ongoing collaboration yields benefits such as shared learning and continuous optimisation of the solar panel system, thereby increasing revenues. This learning process often leads to new opportunities, known as real options, which represent future value.

As noted, farmland can accommodate solar panels constructed above it, and conversely, panels can facilitate agricultural use of the land beneath them if erected at a slightly elevated level. Similarly, car parks can support overhead solar panel installations, with the added option of charging electric vehicles. The real option here involves broadening the scope of applications. Moreover, scaling up operations represents another form of real option, offering future value. Through collaboration, farmers and energy companies can pursue expansion initiatives.

In sum, the economic case for integration—or economic trump card—is supported by four key arguments: reduced costs, enhanced yields, decreased risk, and lower transaction costs coupled with the creation of real options that offset those transaction costs.

The Political Trump Card

Integrality represents the aim of addressing complex issues holistically, acknowledging their inherent interconnectedness. This approach stands in contrast to the traditional sectoral methodology, which tackles problems and generates interventions from the perspective of individual disciplines or "silos." In practice, effective policymaking often necessitates a balance between integrated and sectoral approaches. The timing of implementing an integrated approach is crucial; initiating too early in the policy-making process may result in an overwhelming number of options for coherence, while starting too late may find sectoral choices and budgets already predetermined.

An integrated policy framework facilitates coordination among various stakeholders, encouraging them to align their current activities and explore realistic options for future endeavours. This collaborative approach enables mutual reinforcement, leading to optimisation for all parties involved. It is important to distinguish this from maximisation, where each entity strives to achieve the most for itself without adapting to the collective optimum. While maximisation dominates current economic thinking, integration offers an alternative paradigm: optimisation through synergistic cooperation.

The government is often a key participant in integrated solutions. However, when it does not participate (and sometimes also when it does), other beneficiaries frequently cite internal governmental divisions as significant obstacles. Integration typically transcends departmental boundaries, necessitating the resolution of budget allocation issues from various sources into a single project. This process can be challenging for government bodies unaccustomed to such practices, potentially leading to complications with audits by accountants and parliamentary oversight. Despite these challenges, governmental entities frequently engage in integrated projects.

In the private sector, companies typically operate with a unified budget: all financial flows originate from and return to a single source. Profits are not retained by individual departments, and investments are shared across the organisation. While governments originally operated similarly, political developments have led to the establishment of separate budgets (source: Interviews NL cities). Within governmental structures, the non-intervention principle tends to prevail. Municipal departments maintain autonomy over their work and budgets, with surpluses typically not reverting to a shared budget and investments sourced from departmental allocations. This principle is closely associated with departmental silos. While proponents of integrated working often advocate for decompartmentalisation, it is crucial to define this concept accurately. The existence of departmental partitions allows for a case-by-case assessment of whether sectoral or integrated approaches are more appropriate.

The City of Utrecht exemplifies a coordinated approach to budget allocation, raising funds for the Transwijk and Kanaleneiland neighbourhoods from various budgets. This method emphasises the importance of maintaining scale while encouraging individual budget holders to optimise their spending through coordination with others. Rather than complete decompartmentalisation, this approach invites participation in an investment platform while maintaining existing departmental structures (source: Interviews NL cities).

Integration implies a shift in means rather than ends. For instance, green spaces may serve multiple purposes beyond supporting biodiversity, such as water collection or air purification. This multifunctionality does not negate the original goals but may alter the means of achieving them. Decompartmentalisation, in this context, refers to different parties serving their respective goals through collaborative resource deployment. This collaboration occurs from within their specialised domains, underscoring that integration does not preclude specialisation. This approach facilitates accountability within individual domains.

Political and administrative flexibility is essential to enable the use of resources for multiple purposes. This may involve coordinated use of multiple budgets, such as allocating sewer funds to green roofs or welfare funds to water squares due to their positive impact on quality of life. While this aligns with the concept of a collective governmental budget, it also acknowledges the existence of earmarked funds and demonstrates that such allocations do not preclude coordinated utilisation.

In sum, the political implications of integration—the political trump card—highlight potential benefits while also acknowledging challenges related to financial accountability within governmental structures. This tension underscores the need for a nuanced approach to integrated policymaking, balancing the advantages of holistic problem-solving with the requirements of fiscal responsibility and departmental expertise.

5 Conclusion: The Interplay of Decision-Making Factors in Integrated Solutions

The primary focus of this research is to create more insight into how public and private entities in the Netherlands currently invest in integrated projects and the barriers they encounter. We found that traditional financial instruments, such as loans, are typically designed to support the projects of a single beneficiary by leveraging its capabilities. However, the integration of multiple stakeholders in financing arrangements requires new forms of financing and shared benefits. To meet this need, investment platforms and funds are emerging as mechanisms for organising and optimising financial flows, rather than as financial instruments themselves.

Investment platforms and funds facilitate the cooperation and large-scale investment required for integration by using different financial instruments such as loans, equity, and grants. These instruments open up new channels of funding and foster collaboration between multiple stakeholders, thereby supporting a more integrated approach to project financing. In doing so, they create opportunities to leverage private sector expertise and resources to achieve broader public policy objectives and increase the effectiveness of financial instruments in achieving sustainable cities.

By employing the theory of trumping, the economic and financial benefits of integration became evident, along with the financial tools and methods to achieve these benefits. Analysing the financial trump card, we found that the investment platform in the Cleantech case study reveals several key opportunities for improving project financing and implementation. Overall, the integration of financial mechanisms can significantly improve the viability and attractiveness of integrated investments.

The economic advantages of integration, as shown by assessing the economic trump card and demonstrated by the examples of green roofs and solar panels, are primarily driven by four key arguments: reduced costs, increased yields, reduced risk, and lower transaction costs with the creation of real options. These economic benefits justify public and private investment in integrated business cases. These business cases are unique in that successful investments not only yield higher returns for the initiator but also for other potential co-investors. Still, the Cleantech case study highlights how, despite the strength of economic and financial arguments, these factors may not always dominate the decision-making hierarchy.

Focussing on the political trump card revealed that the compartmentalised nature of public budgets poses a significant barrier, requiring political cooperation for budget flexibility. Participants in an integrated business case understand the necessity of collaboration and perform sub-optimally alone. Horizontal coordination with other parties is optimal, as maximising one's part of the business case does not yield better results. Governments, when involved, may prefer horizontal solutions over vertical ones. Vertical coordination implies the government acts independently, dictating terms to other parties. While a government may justifiably choose vertical solutions, Scott (2020) notes that the gravity of vertical solutions can sometimes overshadow the benefits of horizontal optimisation, even when the latter offers superior financial and economic outcomes.

The quest for efficiency between main effects and co-benefits can lead to diffuse goal prioritisation, causing uncertainty about financing. A clear distinction between the main effects and side effects, co-benefits, and prioritisation of these, facilitates easier financing organisation. Otherwise, integration can result in dependencies that stagnate the process and fragmented responsibility. This leads to confusion about the beneficiaries' identities, cost bearers, and the appropriate financial distribution. This conclusion, supported by Natuurverdubbelaars (2024), explains why governments might revert to vertical coordination.

The proposed theoretical framework has significant implications for governance and financial strategies in integrated projects. It suggests that to successfully implement integrated approaches, stakeholders must:

1. Develop compelling narratives around new efficiencies and multiple value creation.
2. Design or identify the already existing appropriate financial mechanisms that can accommodate and incentivise multi-purpose actions.
3. Recognise the inherent tension between vertical and horizontal coordination.

As Lenferink et al. (2017) note, there is a growing effort to integrate sectors in spatial development, aligning with the principles outlined in this framework. However, the success of these efforts will depend on the ability to overcome the entrenched vertical administrative culture governing expenditure.

In conclusion, this theoretical framework provides a tentative understanding of the challenges and opportunities in implementing integrated governance approaches. Combining insights from systems theory, action theory, and our proposed theory of trumping offers a foundation for analysing and addressing the complexities of integration in public sector management for sustainable cities and finance.

6 Discussion and Future Research

This study shows that a significant barrier to integrated projects is the compartmentalised nature of budgets, which limit the flexibility needed for comprehensive initiatives. Political cooperation is essential to make budgets more versatile. For example, an alderman might need to justify using funds allocated for greenery to purify water. This would only be feasible if the greenery's quality improves or if the organisation responsible for clean water partially funds the initiative. This necessitates the amalgamation and disentanglement of budgets, raising the question of whether municipal councils have patience and trust for such processes. Councils are often protective of their budgets and may resist reallocating funds earmarked for specific purposes to projects with separate budgets.

Strict adherence to compartmentalised spending can be justified in certain contexts. For instance, it is challenging to justify using funds to support incomes for a heating network, although this network has the ability to protect people from energy poverty. Investing in a heating network to prevent energy poverty is a valid argument. Thus, it is crucial for politicians to determine which argument holds the highest priority, which is the trump card.

In the Cleantech case, however, we have found that despite the strength of economic and financial arguments, these factors may not always dominate the decision-making hierarchy. Why this seems to be the case is difficult to explain. One assumption is that the economic and financial trump card can be verified, while the political trump card holds the power to be true regardless of verification. Political decisions are often made in a "black box" to balance transparency with the need for confidentiality and efficiency in governance. This metaphorical black box refers to the opaque processes that take place behind closed doors as policymakers deliberate, negotiate, and reach decisions. Such secrecy is sometimes necessary to protect sensitive information, encourage frank discussions, and prevent premature public scrutiny that could derail complex negotiations. While we acknowledge that the black box approach ensures that certain decisions are made with the necessary depth and care, verifiability is important for making equitable trade-offs between different trump cards. More research is needed on how to achieve transparency in the political decision-making process so that the political trump card can be verified like any other trump card.

Additionally, we acknowledge the existence of more variants of co-investment structures which were not covered in our analysis. A broader assessment of the extent to which these other types of co-investment instruments, like crowdfunding or private equity funds, are able to address the governance and political hurdles exposed in this research, could be worthwhile. Furthermore, a next avenue of research could determine the action perspective of the different types of stakeholders impacted or necessary for such co-investment, e.g. in the face of the pervasiveness of the political trump card, how can both public and private partners act to ease collaborative investment for their mutual benefit?

Finally, the theory of trumping provides a preliminary understanding of the challenges and opportunities in implementing integrated governance approaches. However, in this study, the theory of trumping was only applied through three trump cards: the financial, economic, and political trump cards. In the case studies used for this study, we found that these trump cards predominated and allowed for a better understanding of dynamics in decision-making processes and outcomes. We acknowledge that there are other trump cards that are present in systems and can influence decision-making processes, such as legal trump cards, societal trump cards, and organisational (structural) trump cards. For future research, we recommend identifying and testing additional trump cards. In this way, we can better understand decision-making processes for integrated solutions and create new opportunities for achieving sustainable cities.

Declaration of Competing Interests This research is based for a large part, especially the parts on finance, on work funded by the European Union, ATELIER | Positive Energy Districts—GA No. 864374.

References

Andrews P (2022) Breaking government silos through holistic service integration. Apolitical, Government Trends 2022. https://apolitical.co/solution-articles/en/breaking-government-silos-through-holistic-service-integration

Boström M, Karlsson M (2013) Responsible procurement, complex product chains and the integration of vertical and horizontal governance. Environ Policy Gov 23(6):381–394. https://doi.org/10.1002/eet.1626

Cohen-Shacham E, Walters G, Janzen C, Maginnis S (2016) Nature-based solutions to address global societal challenges. https://doi.org/10.2305/iucn.ch.2016.13.en

Cumiskey L, Priest SJ, Klijn F, Juntti, M. (2019). A framework to assess integration in flood risk management: implications for governance, policy, and practice. Ecol Soc 24(4). https://doi.org/10.5751/ES-11298-240417

De Natuurverdubbelaars (2024) Investeren in integraliteit, Integrale fondsvorming voor meervoudige waardecreatie in de openbare ruimte. City Deal Openbare Ruimte. https://www.citydealopenbareruimte.nl/kennis+en+instrumenten/financiering/2626227.aspx?t=Rapport-Integrale-fondsvorming-voor-waardecreatie

Dorst H, van der Jagt A, Runhaar H, Raven R (2021) Structural conditions for the wider uptake of urban nature-based solutions—a conceptual framework. Cities 116:103283. https://doi.org/10.1016/J.CITIES.2021.103283

Droste N, Schröter-Schlaack C, Hansjürgens B, Zimmermann H (2017) Implementing nature-based solutions in urban areas: financing and governance aspects. In: Theory and practice of urban sustainability transitions. Springer, Cham, pp 307–321. https://doi.org/10.1007/978-3-319-56091-5_18

Duivenvoorden E, Hartmann T, Brinkhuijsen M, Hesselmans T (2021) Managing public space–a blind spot of urban planning and design. Cities 109:103032. https://doi.org/10.1016/j.cities.2020.103032

Floater G, Heeckt C, Ulterino M, Mackie L, Rode P, Bhardwaj A, Carvalho M, Gill D, Bailey T, Huxley R (2016) Co-benefits of urban climate action : a framework for cities. LSE, London. http://eprints.lse.ac.uk/id/eprint/68876

Fuenfschilling L, Truffer B (2014) The structuration of socio-technical regimes - Conceptual foundations from institutional theory. Res Policy 43(4):772–791. https://doi.org/10.1016/j.respol.2013.10.010

Geels FW (2011) The multi-level perspective on sustainability transitions: responses to seven criticisms. Environ Innov Soc Trans 1(1):24–40. https://doi.org/10.1016/j.eist.2011.02.002

Lenferink S, van Geet M, Arts J, Leendertse W (2017) Afstemming N0VI, POVI & MIRT Fase 2—Bevindingen en conclusies. Universiteit van Groningen, Groningen

NetZeroCities (2024) NetZeroCities—about. https://netzerocities.eu/the-nzc-project/

Mazzucato M (2019) Governing missions in the European Union. Independent Expert Rep Res Innov. https://doi.org/10.2777/014023

Puts H, Van der Heijden J (2017) Toekomstwaarde als basis voor meervoudige investeringen—Kosten delen en extra inkomsten genereren. In: Bouma G (ed) Bijdragen aan Plandag 2017, Stichting Planologische Discussiedagen, Antwerpen, pag, pp 187–196

Revi A, Satterthwaite D, Aragon-Durand F, Corfee-Morlot J, Solecki W (2014) Urban areas in climate change 2014: impacts, adaptation, and vulnerability. Part A: global and sectoral aspects. Contribution of Working Group II to the Fifth Assessment Report of the Intergovernmental Panel on Climate Change (Issue January). Cambridge University Press

RLI (2018) De stad als gezonde habitat. RLI, Den Haag

Scott I (2020) Governing by silos. In: Oxford research encyclopedia of politics. Oxford University Press. https://doi.org/10.1093/acrefore/9780190228637.013.1414

Scott I, Gong T (2021) Coordinating government silos: challenges and opportunities. Glob Public Policy Govern 1(1):20–38. https://doi.org/10.1007/S43508-021-00004-Z/TABLES/2

Stipo and DML (2022) Financiering van integratie. City Deal Openbare Ruimte. https://openresearch.amsterdam/nl/page/94628/financiering-van-integratie-openbare-ruimte
Stout M, Love JM (2018) Integrative governance: generating sustainable responses to global crises. https://doi.org/10.4324/9781315526294
STOWA (2018) Facts and values Groenblauwe Daken. https://www.stowa.nl/sites/default/files/assets/PUBLICATIES/Publicaties%20diversen/20180921%20Factsheet%20Groenblauwe%20daken%20GDGD%20sept%202018.pdf. Accessed at 24 Sept 2024
Termeer JAM, Metze TAP (2019) More than peanuts: transformation towards a circular economy through a small-wins governance framework. J Clean Prod 240:118272. https://doi.org/10.1016/j.jclepro.2019.118272
Toxopeus H, Polzin F (2021) Reviewing financing barriers and strategies for urban nature-based solutions. J Environ Manage 289:112371. https://doi.org/10.1016/j.jenvman.2021.112371
Van der Heijden J, Vedder K (2024) De beste energie op de beste plek, Anders investeren in de energietransitie. TNO, Den Haag

Charting Equitable and Sustainable Communities

Mahesti Okitasari, Paul Hudson, and Masachika Suzuki

1 Harnessing Novel Financing and Disruptive Technologies

The transformative potential of novel financing and disruptive technologies can reshape the landscape of urban development. A building design approach, such as patios, offers passive cooling systems rooted in nature-based processes like evapotranspiration and water features and deep cultural and social value in Mediterranean societies, acting as communal hubs that foster interaction and connection to nature (Delgado-Serrano et al., Chap. 2). Collective-based instruments like housing cooperatives present a promising mechanism for addressing both climate change mitigation and adaptation by embedding principles of shared ownership, social responsibility and long-term sustainability at the community level (Akanji and Hudson, Chap. 4). Likewise, fintech applications—ranging from financial services to digital currencies, fraud detection and energy optimisation—contribute uniquely to addressing urban challenges while aligning with the goals of sustainability and inclusivity. There are instances where fintech, such as blockchain-driven energy trading platforms, can integrate renewable energy into city-level resource management (Adi and Aljunaidy, Chap. 9).

When an established incumbent already exists, disruptive innovations need to consider special entrance strategies—in this edited volume, this could be accessibility and affordability. This, for example, might apply to fintech in both rural and urban settings where disrupting traditional finance systems is instrumental in reducing financial exclusion. Adi and Aljunaidy (Chap. 9) argued that by integrating

M. Okitasari (✉)
United Nations University Institute for the Advanced Study of Sustainability, Tokyo, Japan
e-mail: okitasari@unu.edu

P. Hudson
Department of Environment and Geography, University of York, York, UK

M. Suzuki
Graduate School of Global Environmental Studies, Sophia University, Tokyo, Japan

M. Okitasari et al. (eds.), *Equitable and Sustainable Cities*, Advances in 21st Century Human Settlements, https://doi.org/10.1007/978-981-92-2793-8_12

digital technologies, fintech reshapes financial models to broaden access to critical services for underserved populations—mobile money systems support rural communities, while blockchain solutions decrease remittance expenses. Blockchain can also be considered to help secure property rights, and smart monitoring systems could further support the viability and effectiveness of housing cooperatives in climate action. Similarly, Insurtech integrates big data and AI to optimise insurance processes, lowering transaction costs tied to monitoring and claim management (Hudson, Chap. 10). There are potential developments where AI-driven tools such as chatbots could be linked to large language models to automate claims and customer management, while advanced analytics improve risk assessment by addressing uncertainties in natural hazards and behavioural factors (Hudson, Chap. 9). By disrupting existing systems, a synergistic relationship between disruptive technologies and access to finance could significantly impact the service accessibility and affordability.

There is evidence that emerging sustainability niches around novel financing and disruptive technologies have generated sophisticated mechanisms based on existing—formal or informal—structures to advance sustainable cities. For example, housing cooperative structures based on democratic principles can facilitate pooled financial resources for climate-resilient infrastructure and recovery financing (Akanji and Hudson, Chap. 4). Furthering cooperative linkages with energy cooperatives can enhance access to sustainable energy solutions if equipped with strengthened institutional memory, social capital, and financial access, bolstering housing cooperatives' efforts to contribute to climate adaptation and resilience-building (Akanji and Hudson, Chap. 4). Another example is how, by leveraging geothermal and hydroelectric energy to power blockchain activities, fintech establishes a synergistic relationship between financial innovation and environmental stewardship, promoting urban sustainability goals (Adi and Aljunaidy, Chap. 9). In parallel, Van der Heijden et al. (Chap. 11) found that while traditional financial instruments are typically designed to support the projects of a single beneficiary by leveraging their capabilities, investment platforms and funds are emerging as mechanisms for integrating multiple stakeholders in financing arrangements. These instruments open new funding channels and foster collaboration between various stakeholders, thereby supporting a more integrated approach to project financing, leveraging private sector expertise and resources to achieve broader sustainable public policy objectives.

A favourable outcome of deploying disruptive innovations often hinges on policies, incentives and alignment with existing practices. For example, the success of distributed solar energy in the United States relies on the Investment Tax Credit, underscoring the critical role of economic mechanisms in advancing renewable energy adoption (Zhu and Suzuki, Chap. 8). The cases of Jiangxi and Guangdong in China and Vancouver in Canada show that robust policy frameworks and green certifications significantly enhance market confidence and transparency, providing clear guidelines for project selection, management of funds and environmental impact reporting, thereby attracting investors and reducing risk in green and municipal bonds (Liu et al., Chap. 7). On MSW management, Li et al. (Chap. 5) highlighted that incineration technologies and homestead policy innovations need to be embedded into the 'pilot-park-port' complex to achieve cross-level visions from region to companies at national and village levels. Their work in the Jing-Jin-Ji region shows that such a method can divert the relocation project from the traditional demolition resettlement practices, including 'land acquisition (*zhengdi*)' and 'transfer to non-farmer identity (*nongzhuanfei*)' (Li et al., Chap. 5). Moreover, applying or harmonising indigenous knowledge and local knowledge can help promote CSA. Nishi (Chap. 6) found that some of this knowledge, such as farmers' ecological knowledge related to the traditional rice varieties and the local community's experiential knowledge of ecosystem restoration, can promote CSA and create effects on triple-win goals such as developing drought tolerance, flood resistance and promoting energy-efficient agriculture. Appropriately pairing indigenous knowledge and local knowledge with modern science and technologies appears to be effective for CSA despite the origins of initiatives and their specific local contexts, as shown in Wayanad District in Kerala and Zhongliao community of Tamsui District in New Taipei City (Nishi, Chap. 6). Across these examples, community mobilisation and consultation are the critical steps in developing interventions through the identification of problems, knowledge gaps and building consensus, especially when state-led institutions are of little help to niche pioneers.

However, implementing novel financing and disruptive technologies reveals persistent socio-economic disparities in delivering sustainable urban development solutions. The cultural and climatic significance and potential of patios as sustainable and equitable urban solutions, as explored in this volume by Delgado-Serrano, Redaelli and Piedrahita, are often tempered by socio-economic barriers where low-income residents rely on concrete courtyards lacking vegetation, exacerbating risks of heat stress (Delgado-Serrano et al., Chap. 2). Patios may be linked to socio-economic status, signalling an area where urban planning could address inequity in access to beneficial green spaces. Similarly, energy poverty in small towns, driven by reliance on more expensive and less efficient fossil fuels, such as gasoline and kerosene, underscores unequal access to cost-effective energy sources compared to urban areas with centralised utility networks (Zhu et al., Chap. 3). Zhu et al.'s study further illustrates the linkages between the living environment and the people's perception of living and energy expenses, where energy poverty could moderate well-being, and the necessity to understand that energy poverty cannot be solely determined by expenditure, but by looking at energy affordability. Moreover, while housing cooperatives in rapidly urbanising countries like Nigeria offer a model for equitable housing, their ability to self-regulate building practices, encourage behavioural change and promote energy efficiency requires inclusive policies to ensure it benefits lower-income groups (Akanji and Hudson, Chap. 4). These examples reveal that access to urban services is often tied to socio-economic status, perpetuating urban inequities.

2 The Way Forward

Reflecting on the evidence and collective experience presented in this volume, we propose the following priority areas for research to advance the deployment of novel financing and disruptive technologies for equitable and sustainable cities.

Enabling and Managing Disruptions. Multiple contributions in this volume pointed out the numerous different entry points and the wide range of possible disruptions brought by introducing financing and technology innovations. The findings also suggest that novel financing and disruptive technologies affect sectors and stakeholders in varied and complex ways. These incite further efforts to establish a stronger theoretical grounding and sound empirical studies to provide a dynamic understanding of the phenomenon of disruption and how to enable and manage it across phases spanning from entry to mainstream.

Placing People at the Centre of Disruptiveness and Performance. Novel financing and disruptive technology can improve people's quality of life, reduce environmental impact and create more sustainable, equitable and resilient urban environments (Munawar et al. 2022; Sovacool et al. 2022; Yigitcanlar et al. 2021; Zhang et al. 2024). Further study on the perspective of justice and equality is necessary to enrich scholarship and to better understand the linkages between disruptive and novel innovations and inequalities. The utilisation of disruptive technologies and novel financing, such as AI-driven tools and Insurtech, has ethical and risk implications that need to be better understood to ensure that these innovations do not exacerbate inequalities and we keep improvements in overall human welfare at the heart of technological development.

Moderating the Role of Institutional and Legal Environment. Novel financing and disruptive technologies potentially offer versatile and scalable solutions for sustainable cities, and enabling ecosystems, framework and governance play critical roles in mediating their adoption and operationalisation (Liu et al. 2024; Zeng and Jiang 2025). More studies are needed on topics such as how institutional reform or legal environment may accelerate or limit disruptive innovations, what types of supportive regulatory frameworks and multilevel and multisectoral governance facilitate the just and ethical adoption of novel financing and disruptive technology, and which policies encourage partnerships and collaboration.

References

Liu X, Wang C, Zhao S, Ding J, Jia Y (2024) Role of Fintech adoption in the impact of sustainable policy intervention on enterprise transformation in resource-based cities: evidence from China. Resour Policy 88:104443. https://doi.org/10.1016/j.resourpol.2023.104443

Munawar HS, Mojtahedi M, Hammad AWA, Kouzani A, Mahmud MAP (2022) Disruptive technologies as a solution for disaster risk management: a review. Sci Total Environ 806:151351. https://doi.org/10.1016/j.scitotenv.2021.151351

Sovacool BK, Newell P, Carley S, Fanzo J (2022) Equity, technological innovation and sustainable behaviour in a low-carbon future. Nat Hum Behav 6(3):326–337. https://doi.org/10.1038/s41562-021-01257-8

Yigitcanlar T, Mehmood R, Corchado JM (2021) Green artificial intelligence: towards an efficient, sustainable and equitable technology for smart cities and futures. Sustainability 13(16):8952. https://doi.org/10.3390/su13168952

Zeng T, Jiang Y (2025) Fintech and low-carbon urban development. Financ Res Lett 79:107229. https://doi.org/10.1016/j.frl.2025.107229

Zhang W, Wang J, Jin H (2024) Digital finance, innovation transformation, and resilient city growth. Sci Rep 14(1):7056. https://doi.org/10.1038/s41598-024-56998-z

FSC
www.fsc.org
MIX
Papier | Fördert
gute Waldnutzung
FSC® C083411

Kolibri 360 GmbH
Ferdinand-Jühlke-Straße 7
99095 Erfurt, Deutschland
produktsicherheit@kolibri360.de